Teach You to Write

Business E-mails

赖世雄 著

职场E-mail英语

图书在版编目（CIP）数据

教你写职场 E-mail 英语 / 赖世雄著. -- 北京 : 外文出版社，2019.6
ISBN 978-7-119-12066-9

Ⅰ. ①教… Ⅱ. ①赖… Ⅲ. ①电子邮件－英语－写作－自学参考资料 Ⅳ. ① TP393.098 ② H315

中国版本图书馆 CIP 数据核字（2019）第 127871 号

选题策划：知语文化
特约编辑：严　妙 郑新东
责任编辑：李春英
装帧设计：朱月仙
印刷监制：秦　蒙

教你写职场 E-mail 英语

作　者：赖世雄

© 外文出版社有限责任公司

出 版 人：徐　步
出版发行：外文出版社有限责任公司
地　址：中国北京西城区百万庄大街24号　邮政编码：100037
网　址：http://www.flp.com.cn　电子邮箱：flp@cipg.org.cn
电　话：008610-68320579（总编室）　008610-68995964/68995883（编辑部）
008610-68995852（发行部）　008610-68996183（投稿电话）
印　制：北京华创印务有限公司
经　销：新华书店/外文书店
开　本：710 mm × 1000 mm　1/16
印　张：15.5
字　数：430千字
版　次：2019 年 10 月第 1 版第 1 次印刷
书　号：ISBN 978-7-119-12066-9
定　价：38.00 元（平装）

版权所有 侵权必究

国际贸易往来日益频繁，现在电脑、手机都已成为工作中不可或缺的工具。如何使用正确的用词撰写英文电子邮件，以有效地进行员工之间或与客户的沟通，成了现在所有在职人士必备的技能。

本书依据中国人在职场上的需求，以深入浅出的写作方式来编撰。全书共20章50个单元，每章节都是职场人士必备主题："简介""会议""商务旅行计划""建议""邀请""询问""下单""付款""投诉""查询进度""提议""报告""社交场合""应聘工作""协商""财务事项""销售信函""营销""要求澄清""技术"。

本书主要结构：

1. "基本结构"：让读者了解该类主题文章的基本结构及写作技巧。

2. "写作句型"：归纳该类主题常用的结构及句子。

3. "电子邮件范例"：各大主题的商务情境范文。

4. "重要单词短语"：针对文章重要单词短语完全解析，不留任何疑惑。

5. "商务写作句型中译英练习"：让读者阅读完该章节后，立刻测验自己了解了多少。

本书去芜存菁，摒除内容艰涩或过时的商务书信，转用简单清楚且正确的用词来撰写电子邮件，让读者轻松理清英语语法、用字遣词、句子结构等写作重要因素。本书另外在光盘内附 40 篇范文以供读者参考。精读本书必能学会商务电子邮件的写作结构，并熟悉常见的职场单词及用语，必在职场上留下让人难以磨灭的好印象。

Preface

Most non-native English speakers say they prefer writing an e-mail in English to a customer or colleague over speaking face-to-face with him or her. This is understandable: a writer has more time to think of exactly what to express and precisely how to phrase comments and questions than a speaker does. Yet, writing e-mails well—especially business correspondence—can be quite challenging. In order to write effectively and impressively, it's crucial to know key phrases and avoid common errors. That is the purpose of this book—to help give you the tools you need to confidently compose good, error-free e-mails.

Design of the Book

This publication features a total of 50 units divided into 20 chapters based on particular functions or themes common to the business world. For example, Chapter 1 focuses on introductions, and has a unit on introducing yourself, another unit on introducing your company, and a third based on introducing a new product. Chapter 14, to take another example, provides a unit dealing with writing a cover letter that you would send with your resume, and it also contains another unit with phrases to include in a follow-up e-mail or letter after a job interview. Among the 20 chapters, topics include arranging meetings, placing orders for products, writing reports, discussing proposals, responding to complaints, negotiating, discussing financial matters, and writing about marketing-related issues, as well as many others.

The units begin with 10 useful phrases related to the theme of the chapter, along with sample sentences that use them in context. These are followed by sample e-mails (sometimes a single sample, other times two shorter ones per unit), to enhance your understanding of how to use the vocabulary.

In addition to offering readers a valuable learning experience, this book can be used as a handy reference, somewhat like a dictionary or thesaurus would be used. That is, when you are searching for how to express yourself in writing (and, in many cases, speaking as well), you can look at the table of contents and find the situations and functions that suit your needs. Not only can you save time by doing this (rather than waste valuable work time thinking of phrases to use), you can compose your e-mails effectively and confidently by using this book as a guide.

15 种改进电子邮件写作的方式

① Make Your Reader the First Priority 把读者放在首位

Be organized and to the point. Stay focused in your writing and write briefly so you don't waste the reader's time.

要有组织且简短扼要。专注于写作并写得简洁有力，这样才不会浪费读者的时间。

② Write a Good Subject Line 写好标题

Good subject lines are relevant, meaningful and let the recipient know quickly what the e-mail is about. Since people get so many e-mails on a daily basis, they often scan the subject line before they decide whether or not to open them or to simply junk them. Make sure yours doesn't get trashed for lack of a good subject line.

好的标题会切题，富有意义，并可让收件人迅速了解该电子邮件的内容。由于大家每天都会收到非常多的电子邮件，他们通常会快速扫描标题，然后才决定是否要打开该信件或只是将它删除。务必要确定你的电子邮件不会因为缺少好的标题而被对方删除。

③ Use a Standard Greeting 使用标准问候语

Some people consider it a bit rude to ignore the salutation, even in an informal e-mail. Use "Dear" "Hello" or even "Hi" (informal) as a salutation and then the person's name.

有些人认为即便是一封非正式的电子邮件，忽略称呼语仍会显得有些无礼。使用“亲爱的”、“哈喽”、或甚至是“嗨”（非正式）作为称呼语，然后才是该收件人的名字。

4 Specify Who You Are and Why You Are Writing 表明身份及来信理由

This is especially true if you are writing to the person for the first time. Don't make the reader guess who the e-mail is from and your purpose in writing him or her. Confusion is uncomfortable.

尤其如果你是第一次写信给那个人，别让那位读者猜测寄信者是谁及其目的为何。因为困惑会让人感到不舒服。

5 Don't Make People Try to Read Your Mind 不要让人去猜测你的想法

If you are vague and imprecise, the results you receive will very likely be less than satisfactory. You can save time and aggravation (on both your part and the other's part) by being clear. The results will definitely be better if people don't have to guess what you want or mean.

若你用词含糊不精确，结果就很可能会不尽如人意。表达清楚可以节省时间并避免相互激怒（双方都是如此）。假如人们不必猜测你想要什么或你所指为何，结果一定会更好。

6 Avoid Long, Rambling Messages 避免冗长含糊的讯息

Your goal is to be concise, not impress people with your ability to string together ideas. In short, short is good, and long (usually) is bad. If you have many points that are not closely linked, think about splitting them up into different e-mails and sending them separately. Using bullet points also helps if a message is long.

你的目标是简洁，而不是将众多想法串在一起来让人钦佩。总之，简短为上策，冗长（通常）为下策。如果你有很多彼此并非紧密相连的要点，就要考虑将其分成不同的电子邮件分别寄出。假如讯息很长，标准要点同样会有帮助。

7 Choose Between Formal and Casual 在正式与非正式间做选择

Make sure the vocabulary you choose is appropriate to the

situation. Use titles (Mr., Ms., Dr., etc.) in formal correspondence. Avoid slang and emoticons (such as "smilies") unless the e-mails are informal messages between coworkers or friends. In addition, the type of abbreviated vocabulary—e.g. "c u" (see you), "Gr8" (great)—shouldn't be included in formal writing.

务必确保你所选用的词汇适合该情境。在正式书信使用（先生、女士、博士等）的称谓。除非是同事或朋友之间的电子邮件，否则就要避免使用俚语或表情符号（如“笑脸符号”）。此外，缩写词如“c u”（音似 see you 表“再见”），“Gr8”（音似 great 表“太棒了”）不应出现在正式的写作中。

8 Copy and Paste If You Can 尽量使用复制粘贴

If the information you want to send is not too long, consider copying it and then pasting it into your e-mail rather than attaching a document as an attachment. Attachments need to be opened in separate files, such as Word or Adobe Acrobat, and can be a bit time-consuming for the recipient to access. As well, attachments may be worrying for recipients if they don't know the sender very well. In fact, some businesses refuse to accept attachments for fear of being infected by a virus.

如果你要发送的信息不长，便可考虑将其复制，然后贴在电子邮件内，而不是附上附件。附件需要分别用 Word 或 Adobe Acrobat 才能打开，这可能会让收件人在取得资料时花上一些时间。同样地，如果收件人对寄件人不熟的话，附件也许会让他担心。事实上，一些企业因为害怕感染电脑病毒，所以拒绝接受附件。

9 Take the Time to Proofread 要花时间进行校对

Proofreading your e-mail before hitting the send button can save you from making many careless and embarrassing mistakes, such as spelling the person's name incorrectly. It's good to get into the habit of re-reading your e-mails at least once before sending them. Correspondence that is riddled with typos and wrong information creates the impression you are sloppy and perhaps even lazy.

按下发送键之前先校对你的电子邮件，这能让你避免许多粗心及令人难堪的错误，像拼错名字这样的事。养成在发送电子邮件前至少再重读一遍的习惯是件好事。充满拼写错误和错误信息的信件会让人对你产生草率甚至也许是懒惰的印象。

⑩ Watch What You Write 小心你所写的内容

E-mailing may not be as private as you think. If you are at work, the boss might be secretly looking at the e-mails you are sending and receiving. Even if you are e-mailing at home, a hacker might have found a way to hack into your system and take a peek.

发送电子邮件可能不像你想的那么隐密。如果你正在工作，老板可能会偷看你收发的电子邮件。即使你在家中发送电子邮件，电脑黑客也可能会有办法侵入你的系统偷看邮件内容。

⑪ Respect People's Privacy 尊重他人隐私

When sending out a mass e-mail, it's often a good idea to use the BCC (Blind Carbon Copy) function so that others can't see all of the e-mail addresses of the people you sent the e-mail to. For one thing, using the BCC suggests that you are writing an e-mail to just one recipient, rather than firing off the same piece of correspondence to many people, therefore making it less impersonal. For another, some people may not want strangers to know their private e-mail address.

发送大量电子邮件时，使用 BCC（密送）功能通常会是个好主意，如此一来别人看不到其他收件人的地址。一方面，使用密送功能暗示你只是要写给一位收件人，而不是把同样的邮件发送给许多人，这样一来可以让这封电子邮件看起来不会那么没人情味；另一方面，有些人也许不想让陌生人知道他们的私人电子邮件地址。

⑫ Be Careful with Those Buttons 小心那些按键

Don't make the mistake of hitting the "Reply All" button if you only want one person to receive your e-mail. This is especially true if you are criticizing someone who happens to be on that "Reply All" list.

如果你只想要一个人收到你的电子邮件，不要犯下按到“回复所有人”按键的错误。尤其是你批评的那个人刚好在那份“回复所有人”的名单上时。

⑬ Don't Forget to Reply Without Delay 别忘了要迅速回复

Responding promptly sends a message of professionalism, while late replies (or none at all) create the impression you don't care about the recipient. Think how it feels to have an e-mail ignored or to receive an e-mail reply much later than you sent your original message.

迅速回复代表专业，而延迟回复（或完全不回复）会给予对方你不在乎的印象。设想假如你的电子邮件被忽略，或是你原先的讯息过了很久才收到回复，你会有什么感觉？

⑭ Don't Write Anything You Will Regret Later 不要写下任何会让你后悔的文字

Read over your e-mail before you send it, not just for typos but to see if you really mean what you've written. It's usually a bad idea to fire off an angry e-mail while you're still boiling mad. After re-reading your e-mail, you might decide it's better to reword it or, in some cases, cancel it.

在发送你的电子邮件之前要把它看一遍，这不只是为了检查是否有拼写错误，也是要看你所写的是否真的就是你要表达的。当你仍非常愤怒时，发出一封愤怒的电子邮件通常不是个好主意。重新看过你的电子邮件后，你也许会决定最好还是改一下，或者在某些情况下，决定将其删除而不发送出去。

⑮ Manners Matter 态度很重要

Remember that the people whom you send e-mails to may have a different perception of your messages than you do. They may see what you believe to be a simple request as a demanding order, for example. Also, they may see your joke as biting criticism directed at them. The problem is that writing lacks all of the non-verbal clues that exist in face-to-face communication. For that reason, it's important to make sure you carefully and politely phrase your words. Here's another tip: DON'T USE ALL CAPITAL LETTERS. (it looks like you are screaming!)

请记住，你发送电子邮件的收件人对于你的信息可能会有不同看法。举例来说，对你来说简单的请求，他们可能视之为艰巨的要求；对你的笑话，他们可能视之为对他们进行的刺骨批评。问题就在于写信时缺乏一切存在于面对面沟通才有的非语言线索。因此，确保你的措辞谨慎且有礼貌是很重要的。这里有另外一个诀窍：不要全用大写字母。（那看起来很像你在尖叫！）

Contents

电子邮件格式及用语简介

1 Openings and Closings 开头与结尾

The standard way to begin a formal e-mail is by using "Dear (person's name). However, in informal e-mails, it is also common to use "Hello (person's name)" or even "Hi (person's name)". As for the punctuation after the salutation, it varies depending on which country you are from. For example, for formal business letters, Americans and Canadians use a colon: e.g., Dear John:

(In informal e-mails, North Americans sometimes use a comma: e.g., Dear John,) In contrast, the British do not use any punctuation after the salutation: e.g, Dear John

开始写一份正式电子邮件的标准方式是使用“Dear 某人的名字”。然而，在非正式的电子邮件中，使用“Hello 某人的名字”或“Hi 某人的名字”也是很常见的。至于称呼语之后的标点符号，这就取决于你来自哪个国家而有所差别。举例来说，美国人和加拿大人在书写正式的商业书信时会使用冒号。例如：Dear John:

（在非正式的电子邮件中，北美人有时会使用逗号。例如：Dear John,）相对地，英国人在称呼语之后不会使用任何标点符号。例如：Dear John

正式电子邮件称呼语	非正式电子邮件称呼语
Dear John: （美、加人士使用）	Dear John, （北美人士使用）
Dear John （英国人使用）	Dear John （英国人使用）

If you are using a formal title, such as Mr., Ms. or Dr., it must be followed by a person's last name or full name, but never simply by just his or her first name. That is, Mrs. Bolton, Ms. Bolton, Dr. Bolton and Ms. Dorothy Bolton are all correct, but Ms. Dorothy is not. Some people use only a person's first name when writing (e.g., John:). Using just a person's first name provides a sense of seriousness. However, never use just the person's last name.

如果你要使用正式的称呼，诸如 Mr.、Mrs.、Dr.，之后必须接某个人的姓氏或全名，但就是不可以接他们的名字。也就是说，Mrs. Bolton、Ms. Bolton、Ms. Dorothy Bolton 都是正确的，但 Ms. Dorothy 就是错的。有些人写邮件时只写收件人的名字（例如：John:）。只写收件人的名字给人一种严肃的感觉。但是，绝不要只写收件人的姓氏。

称呼的正确用法	称呼的不正确用法
Mrs. / Ms. / Dr. Bolton 或 Mrs. / Ms. / Dr. Dorothy Bolton	Mr. / Ms. / Dr. Dorothy
Dorothy	Bolton

Once you've correctly typed the salutation, you're ready to begin writing the content of the e-mail. One very common method of beginning correspondence is to state the reason you are writing. Some common phrases for doing this are as follows:

一旦你已正确键入了称呼语，你就准备开始写电子邮件的内容了。书信开头的一种很普遍的方式就是陈述你写信的原因。下面列出了一些书信开头的常见用语：

The reason I'm writing (to you) is to...
我来信旨在……

- The reason I'm writing is to inquire about the price of a product.
  我来信旨在询问某产品的价格。

My purpose in writing (to you) is to...
我给你写信的目的是要……

- My purpose in writing to you is to let you know about the upcoming meeting.
  我给你写信的目的是通知你即将召开的会议。

Another common way to begin an e-mail is to mention a specific item or previous event that the person you are writing to is familiar with. For example:

另一种电子邮件的开头方式则是提及某特定事项，或是你的收件人所熟悉的之前的事。例如：

I'm writing with reference to / in regard to / with regard to...

我来信是关于……

- I'm writing in regard to our telephone conversation yesterday.
  我来信是关于我们昨天电话的通话。

I'm writing regarding / concerning...

我来信是关于……

- I'm writing regarding the sales report you sent me recently.
  我来信是关于你最近寄给我的销售报告。

If you are responding to an e-mail that was sent to you, you can begin by expressing a sense of appreciation and warmth.

如果你在回复一封发给你的电子邮件，你可以先表达感激和温暖的感觉。

Thank you (very much) for...

非常感激你……

- Thank you for the e-mail that you sent me yesterday.
  感谢你昨天发给我的电子邮件。

It was good / great to get your e-mail...

能收到你的电子邮件真好……

- It was great to get your e-mail, Brenda.
  布兰达，收到你的电子邮件真好。

Regarding closings, the most common are Sincerely, Yours truly, Best regards, and Regards. Other less used endings include Yours faithfully and Warm (est) / Kind regards, while Cheers is frequently used for very informal messages.

关于结尾敬辞，最常见的就是 Sincerely（敬上）、Yours truly（敬上）、Best regards（诚挚的祝福）、Regards（谨致问候）。其他较少用到的结尾敬辞包括 Yours faithfully（敬上）、Warm(est) / Kind regards（谨致问候），而在

极其非正式的信件中最后常会用到 Cheers（再见）。

Before using these endings, however, it's also convenient to wrap up the content of your letter by using one of the following expressions:

然而，在使用这些结尾敬辞之前，使用下列用语总结信件内容也是很方便的：

I look forward to...
我期待……
- I look forward to discussing this matter in person with you soon.
  我期待能很快与你讨论此事。

If you have any questions, please don't hesitate to...
如果你还有什么问题，请不要犹豫……
- If you have any questions, please don't hesitate to call me.
  假如你还有什么疑问，请不要犹豫打电话给我。

Please let me know if...
请告诉我是否……
- Please let me know if you require any more details.
  请告诉我你是否需要更多细节。

That's all for now...
大概就是这样了……
- Anyway, that's all for now. Take care.
  无论怎样，大概就是这样了。保重。

2 Key Expressions Commonly Used in the Body of E-mails
常用于电子邮件正文中的关键用语

Your choice of which phrases to use in the body of your e-mail depends largely on your purpose for writing. This book contains 50 units designed to cover the majority of scenarios most people face in the business world. Below you will also find some more general expressions that are useful in everyday situations.

在你电子邮件正文当中用语的选择大部分取决于写信的目的。本书包含 50 个单元，设计已大部分涵盖大多数人在商场上会遇到的场景。以下是一些更常见的一般用语，这些用语在日常情况下非常有用。

Giving Good and Bad News 告知好消息和坏消息

I am pleased to inform you (that)... 我很高兴地通知你……

- I am pleased to inform you we have accepted your proposal.
  我很高兴地通知你，我们接受了你的提案。

You will be pleased to know (that)... 你会很高兴得知……

- You will be pleased to know that the shipment has been sent.
  你会很高兴得知货物已经寄出。

We regret to inform you (that)... 我们很遗憾地通知你……

- We regret to inform you that we have decided to choose another supplier.
  我们很遗憾地通知你，我们决定选择另一家供应商。

I'm sorry to tell you that... 我很抱歉地告诉你……

- I'm sorry to tell you that your offer is unacceptable.
  我很抱歉地告诉你，你的报价我们无法接受。

Making Requests 提出请求

I was wondering if... 我在想是否……

- I was wondering if you could assist me in writing the report.
  我在想你是否能帮助我写这份报告。

Is it possible to...? ……可能吗？

- Is it possible to have the goods delivered by Friday?
  这批货是否可能在星期五以前送达？

I would be grateful if... 如果……我会很感激。

- I would be grateful if you could send me a copy of your catalog.
  如果你能寄一份目录给我的话，我将不胜感激。

Offering Help 提供帮助

Would you like me / us to...?
你要我/我们……吗？

- Would you like me to help you with this project?
  你要我帮你一起做这个项目吗？

I / We can (do sth) if you like.
如果你愿意的话我/我们可以（做某事）。

- I can pick you up at the airport if you like.
  如果你愿意的话，我可以开车去机场接妳。

Indicating If You Can Comply or Not　说明是否能遵守指示

Yes, we are able to...　是的，我们能……

- Yes, we are able to meet the deadline.
  是的，我们能赶上最后期限。

Unfortunately, we are unable to...　不幸的是，我们不能……

- Unfortunately, we are unable to comply with your request.
  不幸的是，我们无法遵照你的要求。

③ Dealing with Attachments　处理附件

When attaching files to your e-mail, it's necessary to make reference to it in the content of your correspondence. Here are some simple phrases you can use to do that.

当你把文档附在电子邮件中时，你有必要在邮件内容里提到那个附件。你可以使用下面这些简单用语。

I have attached (sth) to this e-mail.　我已在这封电子邮件中附上（某物）。

- I have attached a copy of the report to this e-mail.
  我已经在此电子邮件中附上一份报告。

Please find (sth) attached.　请查收所附（某物）。

- Please find the expense report attached to this e-mail.
  请查收附在这封电子邮件中的开支报告。

Please see the attachment...　请参阅该附件……

- Please see the attachment for more details of the plan.
  请参阅附件以了解该计划的更多细节。

④ The Tone of the E-mail: Formal or Informal?　电子邮件的语气：正式还是非正式？

In the business world, a lot of correspondence—especially to clients—is formal. Formality is largely achieved in several ways, such as using titles (Mr., Mrs., Dr., etc.), using polite language and avoiding the use of too many contractions. How do you achieve politeness in a letter or an e-mail? Compare the following sentences:

在职场上，许多书信往来都是正式的——与客户的通信尤是如此。正式礼节大多透过一些方式就可做到，诸如使用头衔称呼（Mr.、Mrs.、Dr. 等等）、使用礼貌用语及避免使用太多的缩写形式。你要如何在信件或电子邮件中表现得有礼貌？比较下列句子：

- “不知道你是否可以在这个项目上帮我的忙。”这个句子有两种说法：
  I was wondering if you could help me on this project.（正式）
  Can you help me on this project?（非正式）
- 一般情况下，句子较长且语气委婉的信件视为正式；而句子简短直接的信件则视为非正式。若要表示“不幸地是，我们无法准时交货。”有两种说法：
  Unfortunately, it appears that we are unable to deliver the shipment on time.（正式）
  We can't deliver the shipment on time.（非正式）

Although formality is common in business, there are numerous situations when you want to be informal in e-mails (in some cases, being overly formal can be seen as being unfriendly). When seeking casualness, it's useful to be more direct; however, don't forget to still be polite—you just don't need to be over-polite. One method is to use more contractions. As well, abbreviations, which are discussed in the next section, are another way of gaining informality.

虽然在商场上正式的情况较常见，但是许多情况下你会想在电子邮件中表达非正式的感觉（在某些情况下，过度正式会被视为不友善）。想要随性时，直接一点会有帮助，然而，别忘了还是要有礼貌——就是不需要太礼貌。有一个方法就是多用缩写形式。同样，下一段要讨论的缩略词也能提供非正式的感觉。

5 Acronyms & Abbreviations 缩略词

AGM—annual general meeting 年度股东大会

- Our AGM is scheduled to take place in March.
  我们的年度股东大会预计于 3 月举行。

ASAP—as soon as possible　尽快

• I need that report finished ASAP, please.
拜托，我需要这份报告尽早完成。

BTW—by the way　顺便一提

• BTW, did you get my invitation to the party?
附带一提，你收到我那场派对的邀请函了吗？

B2B—business to business　企业对企业

• The company only does B2B.
这家公司只做企业对企业的生意。

B2C—business to customer　企业对顾客

• Our B2C business is increasing.
我们企业对消费者的业务正在增加。

CC/cc—carbon copy　抄送

• Could you please cc Joanne on this?
请你把这个抄送一份给乔安好吗？

CRM—customer relationship management　客户关系管理

• Ask Jim—he really knows a lot about CRM.
问问吉姆，他真的对客户关系管理懂得很多。

DIY—do it yourself　自己动手制作

• I went to a DIY store over the weekend.
上周末我去了一家自己动手制作的商店。

FAQ—frequently asked questions　常见问题

• Check out our FAQ page for more information.
查看我们的“常见问题”那一页以获得更多信息。

FYI–for your information　供你参考

• FYI, Mary is now in charge of that project.
供你参考，玛丽现在负责这个项目。

KPI—key performance indicators　关键业绩指标

- Make sure to check the KPI.
  请务必查看关键业绩指标。

LOL—laughing out loud (*slang used when making a joke; used for very, very informal e-mails)　放声大笑（开玩笑时用的俚语，用在极不正式的电子邮件中）

- I can't believe you said that. LOL!
  我不敢相信你说了那话。笑死我了！

GM—general manager　总经理

- The GM has called an important meeting for Thursday at 2 p.m.
  总经理召集了一个重要会议，时间是星期四下午两点。

NA (N/A)—not applicable or not available　不适用或不提供

- Sorry, these items are NA right now, Joe.
  抱歉，乔，这些商品我们暂不提供。

PR—public relations　公共关系

- Henry works for a big PR firm in Chicago.
  亨利任职于芝加哥某大型公关公司。

Q&A—question and answer　问与答

- There will be a Q&A session after the speech.
  那个演讲后，会有问答环节。

Re—in reference to　关于

- Re the conference in May, there is still a lot to be done.
  关于 5 月的那个研讨会，还有很多工作要做。

R&D—research and development　研究与开发

- That company spends a lot on R&D.
  那家公司在研发上花了很多钱。

ROI—return on investment　投资回报率

- The investors are looking for ways to increase their ROI.
  那些投资者正在寻找能够提高投资回报率的方式。

RSVP—réspondez s'il vous plaît (French for "please respond")
敬请赐复（法语的答复）

- Please RSVP by October 13.
  请在 10 月13 日前答复。

TBA—to be announced　应予公布

- The date of the meeting is still TBA.
  本次会议的日期即将公布。

VIP—very important person　贵宾

- We've got a few VIPs coming to the office tomorrow.
  明天我们会有一些贵宾来公司。

⑥ Comparison of an E-mail to a Formal Written Letter
电子邮件和正式书写信件的比较

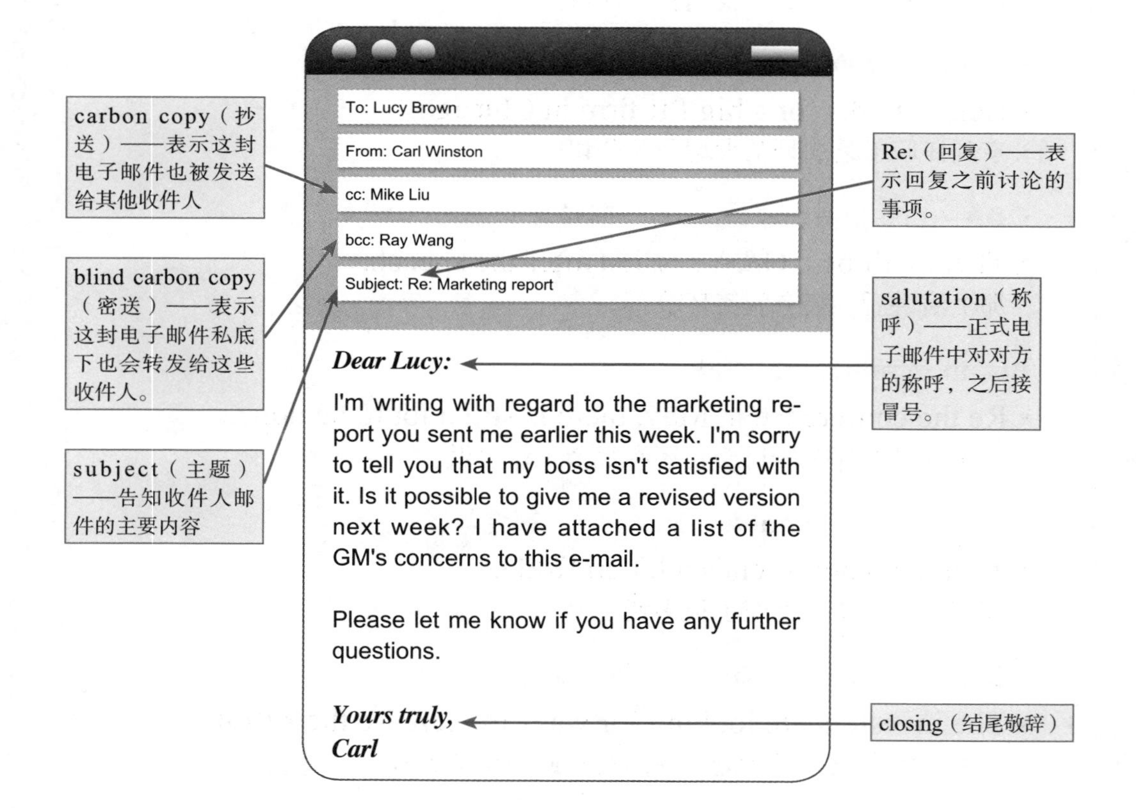

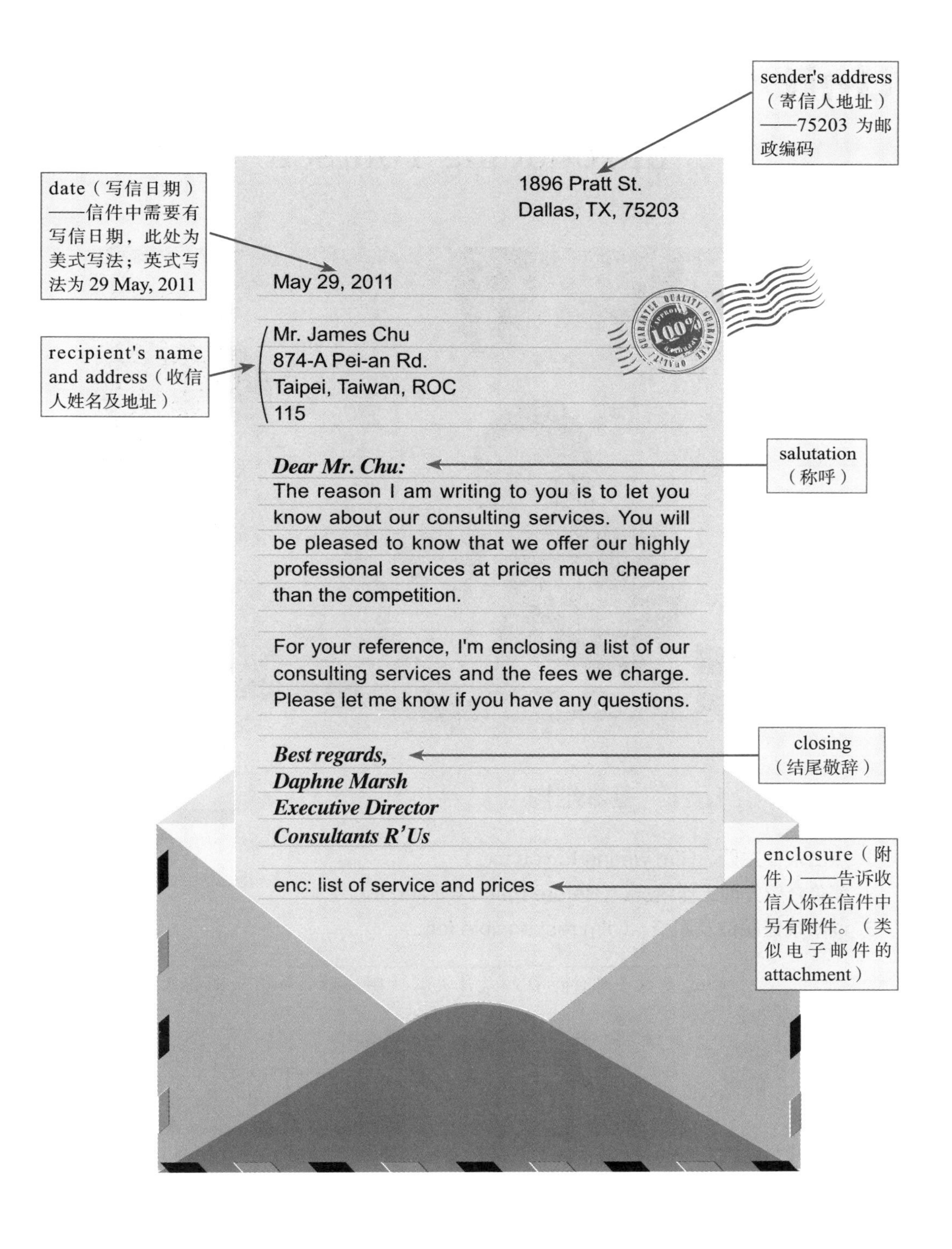
sender's address（寄信人地址）——75203 为邮政编码
1896 Pratt St.
Dallas, TX, 75203
date（写信日期）——信件中需要有写信日期，此处为美式写法；英式写法为 29 May, 2011
May 29, 2011
recipient's name and address（收信人姓名及地址）
Mr. James Chu
874-A Pei-an Rd.
Taipei, Taiwan, ROC
115
salutation（称呼）
Dear Mr. Chu:
The reason I am writing to you is to let you know about our consulting services. You will be pleased to know that we offer our highly professional services at prices much cheaper than the competition.
For your reference, I'm enclosing a list of our consulting services and the fees we charge. Please let me know if you have any questions.
closing（结尾敬辞）
Best regards,
Daphne Marsh
Executive Director
Consultants R'Us
enc: list of service and prices
enclosure（附件）——告诉收信人你在信件中另有附件。（类似电子邮件的 attachment）

Unit 1

Introducing Yourself 自我介绍

Basic Structure 基本结构

1. 点出来信主题。(I'm writing to you to...)
2. 简单介绍先前工作情况。(Before that, I worked for...)
3. 说明现任职位及职责。(I'm responsible for...)

- 这类电子邮件的目的是让大家对刚任职的你有大概的认识，因此内容可包括：
  1. 先前的职称（your previous job title）；
  2. 目前的职称及工作职责（your current position and job duties）。
- 为了有助于往后的合作，可以用"I look forward to meeting you soon."（本人期待能很快与您会面。）和"I'm also eager to sit down with you and hear your ideas."（本人也渴望与您商谈，听取您的想法。）之类的话来向大家请教。
- 不需要在邮件结尾处写上 Thank you（谢谢您），如果内容已很有礼貌，便没有必要使用此类用语。

Sentence Patterns 写作句型 01

I'm writing to you to V./ The reason I'm writing to you is to V.
我来信旨在……

- I'm writing to you to tell you about our company.
  我来信旨在向您介绍一下本公司。
- The reason I'm writing to you is to introduce myself.
  我来信旨在向您做自我介绍。

I joined the company (date / day / time). / I started working for the company (date / day / time).
我于（某日期 / 某天 / 某时间）开始任职于该公司。

- I joined the company at the beginning of this month.
  我于这个月初开始任职于该公司。
- I started working for the company in early May.
  我 5 月初开始任职于该公司。

Previously,... / Before that,...
先 / 之前，……

- Previously, I was a sales representative with a multinational company.
  先前，我在一家跨国公司担任销售代表。
- Before that, I worked for Paramount Products for three years.
  之前，我在派拉蒙产品公司工作了 3 年。

I'm responsible for + N / V-ing / I'm in charge of + N / V-ing
我负责 / 掌管……

- I'm responsible for sales in the Asia-Pacific region.
  我负责亚太区销售业务。
- I'm in charge of the sales department.
  我掌管销售部。

I look forward to + V-ing / I'm eager to V
我期待…… / 我渴望……

- I look forward to meeting you in the future.
  我期待未来能与您碰面。
- I'm eager to discuss this matter with you.
  我渴望与您讨论此事。

Model E-mail 电子邮件范例 01

To: Dean Lansing, project manager

From: Jenny Rockfort

Subject: Self-Introduction

Dear Mr. Lansing:

I'm writing to you and all of my other **colleagues**[1] here at Aim Communications to **introduce**[2] myself. I have recently been **appointed**[3] marketing manager. I joined the company last week. Before that, I was an **assistant**[4] marketing manager for a company named Best Value. As the new marketing manager for Aim Communications, I'm responsible for **developing**[5] new **promotions**[6] to help increase our sales. I look forward to meeting you soon. I'm also eager to sit down with you and hear your ideas on our **upcoming**[7] marketing and advertising **campaigns**[8].

Yours truly,

Jenny Rockfort
Marketing Manager

中译

收件人：项目经理　迪恩・兰辛
发件人：珍妮・罗克福特
主　题：自我介绍

亲爱的兰辛先生：

本人来信旨在向您及目标通信其他同仁自我介绍。本人近日受聘担任营销经理。本人于上星期加入公司。在此之前，本人在一家名为“超值”的公司担任营销副经理。身为目标通信新任营销经理，本人负责开发新的促销活动，以协助提高业绩。本人期待能很快与您会面。本人也渴望与您洽谈，听取您对即将进行的营销及广告活动的想法。

目标通信营销经理
珍妮・罗克福特

Vocabulary and Phrases

1. **colleague** [ˋkɑlig] *n.* 同事
(= co-worker [ˋkoˏwɝkɚ])
2. **introduce** [ˏɪntrəˋd(j)us] *vt.* 介绍
introduce A to B 把 A 介绍给 B
例: Nick introduced me to an American girl at the party.
(尼克在派对上把我介绍给一位美国女孩。)
3. **appoint** [əˋpɔɪnt] *vt.* 任命；委派
appoint sb (as) + 职位
任命某人担任某职位
appoint sb to V
指派某人从事……
例: Brian has just been appointed manager of the sales department.
(布莱恩刚被任命担任销售部经理。)
A detective was appointed to investigate the case.
(某刑警被指派调查此案。)
4. **assistant** [əˋsɪstənt] *a.* 协助的 & *n.* 助理，助手
an assistant manager
副经理(千万不可译成“助理”)
例: My assistant will show you how to operate this machine.
(我的助理将为你示范如何操作这台机器。)
5. **develop** [dɪˋvɛləp] *vt.* 开发，研发
例: The company is spending a large amount of money developing new products.
(该公司正花费大笔资金研发新产品。)
6. **promotion** [prəˋmoʃən] *n.* 促销活动
promote [prəˋmot] *vt.* 促销
例: We need to come up with new ways to promote our products.
(我们需要想出新方法来促销我们的产品。)
7. **upcoming** [ˋʌpˏkʌmɪŋ] *a.* 即将来临的
8. **campaign** [kæmˋpen] *n.* (为社会、商业或政治目的而进行的)运动
例: The government launched a campaign to stop drunk driving.
(政府发起一项遏止酒醉驾驶的运动。)
* launch [lɔntʃ] *vt.* 发起(有组织的活动)

Business Writing Exercises

请按括号中的提示将下列句子译成英文。

1. 我给你写信的理由就是要感谢你为我所做的一切。
____
2. 我在 2008 年的夏天加入本公司。
____
3. 之前，我在一家便利商店打工。(work part-time / be a part-time worker)
____
4. 目前，我负责我们产品的质量管理。(quality control)
____
5. 我很期待未来跟你们合作。(cooperate with sb)
____

Unit 2

Introducing Your Company 介绍公司

Basic Structure 基本结构

1. 表明如何取得联系人信息。(A colleague of mine gave me your name...)
2. 简单介绍公司产品（We produce...）与优点（Our specialty is...）。
3. 说明公司成立历史（We were established...）与地址（We are based in...）。

- 这类电子邮件的目的是要把公司介绍给客户，因此内容可包括：
  1. 公司背景（company background）;
  2. 公司产品特色（product features）。
- 介绍内容须清楚简洁，不用附上许多相关资料。假如对方想知道更多信息，自然会向我方要求提供。
- 在信息中额外加上几个感性的词可以助你建立友好关系。例如：
  I would be happy to meet with you and explain more about our company and products.

Sentence Patterns 写作句型 02

Our company specializes in + N / V-ing / Our specialty is + N / V-ing
本公司专门制造…… / 本公司的专长是……

- Our company specializes in (making) notebook computers.
  本公司专门制造笔记本电脑。
- Our specialty is (making) printers.
  本公司的专长是生产打印机。

We manufacture... / We produce...
本公司生产……

- We manufacture cell phones.
  本公司生产手机。
- We produce power switches.
  本公司生产电源开关。

We offer a full range of... / Our company provides...
本公司供应全系列的…… / 本公司提供……

- We offer a full range of cosmetic products.
  本公司供应全系列的化妆品。
- Our company provides telecommunication services.
  本公司提供电话通讯服务。

Our company was founded in (year) / We were established...
本公司创立于…… / 本公司已成立……

- Our company was founded in 1998.
  本公司创立于 1998 年。
- We were established more than 20 years ago.
  本公司已成立 20 多年。

We are based in... / Our headquarters are located in...
本公司总部设立于…… / 本公司总部位于……

- We are based in Taipei.
  本公司总部设立于台北。
- Our headquarters are located in New York.
  本公司总部位于纽约。

Model E-mail 电子邮件范例 02

To: Ms. Kathleen Simmons, Yellowstone Caterers

From: Derrick Huang, Super Software

Subject: Allow me to introduce my company to you

Dear Ms. Simmons:

A colleague of mine, Raymond Mull, gave me your name and said you might **be interested in**[1] our products. Our company **specializes in**[2] developing software for the food and **beverage**[3] industry. We produce software that is **designed**[4] to make your business **run**[5] more **efficiently**[6]. **In addition**[7], we offer a full range of **inventory**[8] and **accounting**[9] software. Super Software was founded in 2004. Our **headquarters**[10] are located in Hong Kong, and we have **branch**[11] offices in Taipei, Singapore, Kuala Lumpur, Shanghai and Beijing.

I would be happy to **meet with you**[12] and explain more about our company and products.

Sincerely,

Derrick Huang
Sales Representative
Super Software

中译

收件人：黄石酒席承办公司　凯瑟琳・西蒙斯女士
发件人：超级软件公司　德里克・黄
主　题：容我向您介绍本公司

亲爱的西蒙斯女士：

我的同事雷蒙德· 马尔告知我您的大名，并表示您可能对本公司产品感兴趣。本公司专门研发饮食业专用软件。我们生产的软件旨在使贵公司的运营更有效率。此外，我们提供全系列的仓管及会计软件。“超级软件”公司创立于 2004 年。公司总部设立于香港，另外在台北、新加坡、吉隆坡、上海和北京设有分公司。

我很乐意与您安排会面，并针对本公司和产品向您作更多介绍。

超级软件公司销售代表
德里克・黄

Vocabulary and Phrases

1. **be interested in...** 对……感兴趣
= take an interest in...
例: Strange to say, the music teacher isn't interested in music.
（说来也怪，这位音乐老师对音乐不感兴趣。）

2. **specialize** [ˋspɛʃəˏlaɪz] *vi.* 专长
specialize in sth 专门从事某事
例: The shop specializes in handmade cookies.
（那家店专营手工制作饼干。）

3. **beverage** [ˋbɛvərɪdʒ] *n.* 饮料
alcoholic beverages 含酒精饮料

4. **design** [dɪˋzaɪn] *vt.* 设计
be designed to V 被设计用来……
例: This book is designed to teach students how to write.
（设计这本书的目的是教学生如何写作。）

5. **run** [rʌn] *vt.* 经营
(= operate [ˋɑpəˏret])
例: It is Mary's dream to run a clothing business when she finishes college.
（玛丽的愿望就是大学毕业后能经营服装业。）

6. **efficiently** [ɪˋfɪʃəntlɪ] *adv.* 有效率地

7. **In addition,...** 此外，……
例: Jack is handsome. In addition, he is very helpful.
（杰克很英俊。此外，他很乐于助人。）

8. **inventory** [ˋɪnvənˏtorɪ] *n.* 库存

9. **accounting** [əˋkaʊntɪŋ] *n.* 会计

10. **headquarters** [ˋhɛdˏkwɔrtɚz] *n.* 总部（恒用复数）

11. **branch** [bræntʃ] *n.* 分公司；分店
例: The bank has branches all around the world.
（那家银行在世界各地都有分行。）

12. **meet (up) with sb** 与某人相约会面
例: Ned is going to meet with our manager to discuss the project.
（奈德将与经理会面讨论该项目。）

Business Writing Exercises

请按括号中的提示将下列句子译成英文。

1. 本公司专门为小型企业架设网站（build websites）。

____

2. 我们生产各式各样的家具。（all kinds / types of）

____

3. 我们供应全系列的厨房器具。（kitchen utensils）

____

4. 本公司创立于 1997 年 8 月。

____

5. 我们的总部设于东京，但我们有一家分公司（branch office）在台北。

____

Unit 3

Introducing Your Products 介绍产品

Basic Structure 基本结构

1. 点出来信主题。(I am attaching a list of our most popular products.)
2. 简单介绍产品样式(They are available in...)和特点。
3. 推荐最畅销的产品。(I suggest you look at our best-selling...)

- 这类电子邮件的目的是要把产品介绍给客户，因此内容可包括：
  1. 产品样式(product styles)及特点(product features)。
  2. 最畅销的产品(the best-selling product)。
- 要让对方对己方产品有信心，可用以下的句子表示：
  1. Our products are very reliable.(本公司的产品相当可靠。)
  2. Our products are quite durable.(本公司的产品相当耐用。)

Sentence Patterns 写作句型 03

These products come in... / They are available in...
这些产品有……（大小、颜色、样式、材质等）

- These chairs come in four different styles.
  这些椅子有 4 种不同样式。
- They are available in leather, cloth and plastic.
  它们的材质有皮革、布及塑料。

It has many great features, such as... / It also functions as...
这个产品有许多很棒的特色，例如…… / 这个产品也有……的功能

- It has many great features, such as an auto-focus camera.
  这个产品有许多很棒的特色，例如可作为一台自动对焦相机。
- It also functions as an MP3 player.
  这个产品也有 MP3 播放机的功能。

We offer a wide array of... / We have a great selection of...
本公司提供 / 售有各式各样的……

- We offer a wide array of sports equipment.
  本公司售有各式各样的运动器材。
- We have a great selection of garden tools.
  本公司售有各式各样的园艺工具。

Our product line includes... / Included in our product range is / are...
本公司的产品系列包括……

- Our product line includes pet food and toys.
  本公司的产品系列包括宠物食品和玩具。
- Included in our product range are paint, brushes, and wallpaper.
  本公司产品系列包括油漆、刷子和壁纸。

Our products are very reliable. / Our products are quite durable.
本公司的产品相当可靠。/ 本公司的产品相当耐用。

- Our air conditioners are very reliable.
  本公司的空调设备非常可靠。
- Our shoes are quite durable.
  本公司的鞋相当耐穿。

Model E-mail 电子邮件范例 03

To: Keith Barlow

From: Amanda Lee, Furniture Unlimited

Subject: Re: Our products

Dear Mr. Barlow:

Further to[1] our conversation yesterday, I am **attaching**[2] a list of our most popular products. One of our **specialties**[3] is **elegant**[4] sofas that come in many different colors and material. We also have a great **selection**[5] of coffee tables as well as lamps. I suggest you look at page six of the attachment, where you'll find our best-selling lamps, including the XLNT-360. A great feature of this lamp is its ability to **last**[6] for thousands of hours. Please be **assured**[7] our products are **reliable**[8] and **durable**[9]. Don't **hesitate**[10] to contact me if you have any questions about anything in our product range.

Best regards,

Amanda Lee
Customer Accounts Supervisor
Furniture Unlimited

中译

收件人：凯斯·巴洛
发件人：阿曼达·李，无限家具公司
主　题：回复：本公司产品

亲爱的巴洛先生：

针对我们昨日的谈话，兹附上本公司最受欢迎的产品清单。本公司最具特色的产品之一是精致考究的沙发，这些沙发有许多不同的颜色和材质。本公司同时销售各式各样的茶几和台灯。我建议您参看附件第 6 页，内有本公司最畅销的台灯，包括型号 XLNT-360 的一款。这款台灯的特色是能持续照明几千个小时！请放心，本公司的产品可靠且耐用。关于产品系列如有任何疑问，尽管与本人联系。

诚挚的祝福

客户账户主管
阿曼达·李

Vocabulary and Phrases

1. **further to...** 承接；继续（用于信件开头）
   例: Further to our previous e-mail, I'd like to add that our prices are reasonable.
   （承接前一封邮件内容，我想补充一句本公司的价格都很公道。）
2. **attach** [əˋtætʃ] *vt.* 把……附上
   attachment [əˋtætʃmənt] *n.*（电子邮件的）附件
   attach A to B　把 A 附在 B 上
   例: I've attached a photo to my application form.
   （我已把照片附在申请表上。）
3. **specialty** [ˋspɛʃəltɪ] *n.* 专长；特产
4. **elegant** [ˋɛləgənt] *a.* （服饰）优雅的；（陈设）讲究的，精美的
5. **selection** [səˋlɛkʃən] *n.* 可供选择之物（用于下列短语）
   a good / wide selection of...
   各式各样的……
   例: We have a wide selection of patterns for you to choose from.
   （本公司有各式各样的款式供您选择。）
6. **last** [læst] *vi.* 持续
   例: The meeting only lasted (for) half an hour.
   （那次会议只持续了半小时。）
7. **assure** [əˋʃʊr] *vt.* 使确信，使放心
   assure sb + that 从句　向某人保证……
   例: The mechanic assured me that the car would be ready by tomorrow.
   （修车师傅向我保证明天以前车子就会修好。）
8. **reliable** [rɪˋlaɪəbl̩] *a.* 可靠的
   例: We're looking for someone who is reliable and hard-working.
   （我们在找可靠而又勤劳的人。）
9. **durable** [ˋdʊrəbl̩] *a.* 耐用的
10. **hesitate** [ˋhɛzəˏtet] *vt.* 顾虑（之后接 to 引导的不定式短语做宾语）
    don't hesitate to V　不要犹豫……
    例: Don't hesitate to contact me if you have any questions.
    （如有疑问，请尽管与我联系。）
    *contact [ˋkɑntækt] *vt.* 联系

Business Writing Exercises

请按括号中的提示将下列句子译成英文。

1. 这款衬衫有 3 个尺寸，7 种颜色。（This shirt comes in...）

   ____

2. 这台液晶电视（LCD TV）也可当监视器（monitor）用。

   ____

3. 我们销售各式各样的钟表。（a wide array / selection of）

   ____

4. 我想提醒您，我们所有的产品都可靠又耐用。（reliable and durable）

   ____

5. 如果你需要任何帮助，尽管给我打电话。

   ____

Unit 1

Arranging Meetings 安排会议

Basic Structure 基本结构

通知

1. 点出会议主题及讨论事项。（I think we need to meet to discuss the Berlin project.）
2. 说明开会时间。（The meeting is scheduled for...）

回复

1. 确认会议时间并回复是否方便。
2. 如有不便，则询问另一时间是否可行。（Is next Monday or Tuesday all right with you?）

- 在此封电子邮件中，收件人和发件人为同事。如要求跟公司以外的联系人见面时，应说明会面的地点、日期和时间。如果过去和对方没有任何接触，就应在信中说明你如何取得对方的资料，并提供你公司的一些资料。
- 如果你希望出席者在会议上携带某些资料，可以在邮件中提醒他。例　如：Don't forget to bring the report on last month's sales figures to the meeting.（不要忘了把上个月销售数字的报告带到会上。）

Sentence Patterns 写作句型 04

Do you have time... (date / day / time)? / Are you available... (date / day / time)?
您（某日期 / 某天 / 某时间）有空吗？

- Do you have time to meet on Monday or Tuesday afternoon?
  您星期一或星期二下午有空开会吗？
- Are you available on Sept. 27 in the morning?
  您 9 月 27 号早上有空吗？

Is (date / day / time) all right with you? / What's your schedule like (date / day / time)?
您在（某日期 / 某天 / 某时间）方便吗？ / 您在（某日期 / 某天 / 某时间）的日程安排如何？

- Is this Friday at noon all right with you?
  这星期五中午您方便吗？
- What's your schedule like tomorrow?
  您明天的日程安排如何？

Please let me know if (date / day / time) is good for you. / Please respond at your earliest convenience.
请告知（某日期 / 某天 / 某时间）您是否方便。/ 请您尽早回复。

- Please let me know if Wednesday the 12th is good for you.
  请告知我 12 号星期三您是否方便。
- Please respond at your earliest convenience regarding this matter.
  请您尽早回复相关事宜。

Let me check my calendar. / I need to check my schedule.
让我查一下我的日程表。/ 我需要查一下日程安排。

- Let me check my calendar to see if I'm free on that day.
  让我查一下我的日程表看看我当天是否有空。
- I need to check my schedule to see if I'm available.
  我需要查一下我的日程安排看看我是否有空。

I'll get back to you soon about (the event). / I'll let you know ASAP.
我会很快再与您联系相关（事件）。/ 我会尽早通知您。

- I'll get back to you soon about the meeting.
  我会很快再与您联系会议的相关事宜。
- I'll let you know ASAP about this.
  我会尽早让您知道此事。
  * ASAP = as soon as possible 尽快

Model E-mail 电子邮件范例 04

To: Karen Anderson

From: Terry Richards

Subject: Meeting to discuss the project

Hi Karen:

I think we need to meet to discuss the Berlin project. There are still a lot of matters we need to **iron out**[1]. Do you have time on Wednesday or later this week? Please **respond**[2] **at your** earliest **convenience**[3].

Regards,
Terry

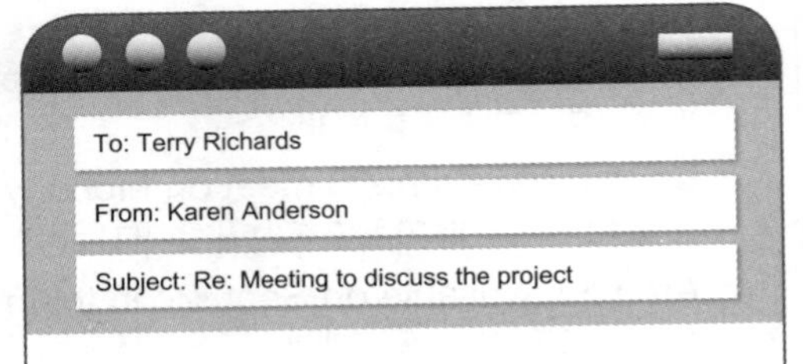

Hello Terry:

I just got your **message**[4]. Unfortunately, I'm **on my way** out of the office **to**[5] visit a **client**[6]. Regarding the meeting, please let me check my calendar before I make a **commitment**[7]. I'll **get back to**[8] you soon about this. If this week is not good for me, is next Monday or Tuesday all right with you? What's your schedule like next week?

Karen

中译

收件人：凯伦·安德森
发件人：泰瑞·理查兹
主　题：开会讨论方案

凯伦：

　　你好！

　　我觉得我们需要开会讨论柏林方案。还有很多问题等着我们解决。你在本周三或之后有空吗？请尽快回复。

　　诚挚的祝福

泰瑞

中译

收件人：泰瑞·理查兹
发件人：凯伦·安德森
主　题：回复：开会讨论方案

泰瑞：

　　你好！

　　我刚刚才收到你的信息。不巧的是，我正离开办公室去拜访客户。有关会议一事，在承诺之前，请先让我查看日程表。我会尽快就此事再与你联系。如果本星期对我不便的话，请问下星期一或星期二你是否方便？你下星期的日程安排如何？

凯伦

Vocabulary and Phrases

1. **iron out sth** 解决某事

   例: We need to iron out how to cut down the budget first.
   （我们必须先解决如何削减预算。）

2. **respond** [rɪˈspɒnd] *vi.* 回应(与介词 to 连用)

   respond to... 回应……

   例: How did Tim respond to the news that his fiancée is pregnant?
   （提姆对他未婚妻怀孕这个消息有何回应？）

3. **at sb's convenience** 某人方便时

   at sb's earliest convenience
   某人尽早

   例: Can you telephone me at your convenience to arrange a meeting?
   （你能不能在方便时给我打个电话，安排见一次面？）

4. **message** [ˈmɛsɪdʒ] *n.* （书面或口头）消息；信息

   leave a message with sb
   留信息由某人转达

   take a message 帮某人留言

   例: If I'm not there, leave a message with Kerry.
   （如果我不在那儿，留信息由凯莉转达。）

5. **on one's way to + 地点名(如: the station / the park)** 某人前往……途中

   on one's way + 地点副词（如：here / there / home） 某人前往……途中

   on one's way out of the office
   某人从办公室出来途中

   例: I dropped in on John on my way home.
   （我回家途中顺便去看约翰。）

   *drop in on sb 顺道拜访某人

6. **client** [ˈklaɪənt] *n.* 客户

7. **commitment** [kəˈmɪtmənt] *n.* 承诺

   make a commitment 做出承诺

   例: I don't want to make a big commitment to Cindy at the moment.
   （我目前不想对辛蒂做出重大承诺。）

8. **get back to sb** 回复某人

   例: I'll get back to you later with those sales figures.
   （我稍后会回复你那些销售数字。）

Business Writing Exercises

请按括号中的提示将下列句子译成英文。

1. 你星期五晚上有空吗？

   ____

2. 你接下来几天的日程安排如何？

   ____

3. 请你在方便时尽早给我发电子邮件。(at your earliest convenience)

   ____

4. 在你帮我订班机（book a flight）之前，让我先查一下我的日程表。

   ____

5. 会议一结束我会马上再跟你联系。

   ____

Unit 2

Agreeing to or Rejecting Proposed Meeting Times 同意或拒绝所提议的会议时间

Basic Structure 基本结构

1. 说明已收到对方通知会议时间的邮件。(I received your e-mail about the series of meetings.)
2. 简单说明可接受(Sure,... is fine with me.)或不能接受的会议时间(Sorry,... is impossible for me.)。
3. 针对不合适的时间，询问对方是否可以改约另一个时间。(Can we make it for another time?)

- 在回复对方的开会要求时，句子应保持简短并点出重点。
- 如果你可以在指定的时间出席，可以使用 Great 或 Sure 等词来表示你赞成这样的安排。例如：
  Great, I'll see you Thursday at 3 p.m. in the boardroom.（太棒了，那就星期四下午 3 点在董事会会议室见。）
- 尽量在同一天回复邮件，以便对方迅速做出进一步的安排。

Sentence Patterns 写作句型 05

Yes, (date / day / time) works for me. / Sure, (date / day / time) is fine with me.
是的，（某日期／某天／某时间）我可以。／当然，（某日期／某天／某时间）我可以。

- Yes, Bill, July 5 works for me.
  好的，比尔。7 月 5 号我可以。
- Sure, Tuesday at 2 p.m. is fine with me.
  当然，星期二下午 2 点我可以。

Great, I'll see you (date / day / time). / That time is perfect. I'll talk to you (date / day / time).
太棒了，那就（某日期／某天／某时间）见。／那个时间非常合适，那就（某日期／某天／某时间）再谈。

- Great, I'll see you Thursday at 3 p.m. in the boardroom.
  太棒了，那就星期四下午 3 点在董事会会议室见。
- That time is perfect. I'll talk to you on Friday.
  那个时间非常合适。那就星期五再谈。

I'm free (date / day / time), but not (date / day / time). / How about (date / day / time), instead?
我（某日期／某天／某时间）有空，但（某日期／某天／某时间）没空。／那么改为（某日期／某天／某时间）如何？

- Steve, I'm free on August the 12th, but not on the 13th.
  史蒂夫，我 8 月12 号有空，但 13 号没空。
- How about Wednesday morning, instead?
  那么改为星期三早上如何？

I'm afraid I'm tied up (date / day / time). / Sorry, (date / day / time) is impossible for me.
恐怕（某日期／某天／某时间）我会很忙。／抱歉，（某日期／某天／某时间）我不行。

- I'm afraid I'm tied up on Thursday the 19th, David.
  大卫，恐怕我 19 号星期四会很忙。
- Sorry, 3 p.m. on Tuesday is impossible for me.
  抱歉，星期二下午 3 点对我来说不行。

Another time is preferable to me. / Can we make it for another time?
另一个时间比较适合我。／我们能不能改约另一个时间？

- Another time is preferable to me. How about, say, Thursday or Friday?
  另一个时间比较适合我。譬如，星期四或星期五如何？
- Can we make it for another time? Perhaps next week would be better.
  我们能不能改约另一个时间？或许下星期会好一点。

Model E-mail 电子邮件范例 05

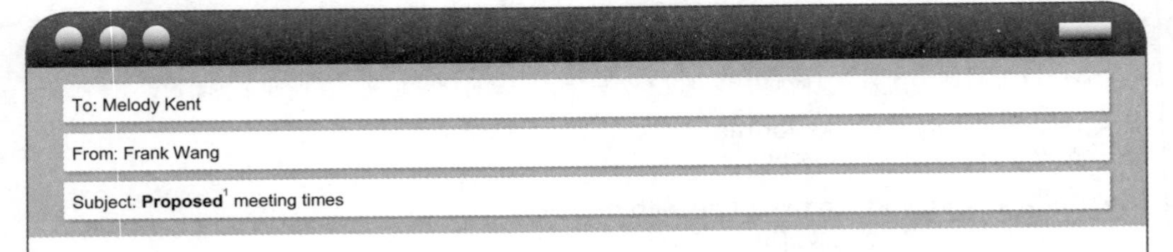

To: Melody Kent

From: Frank Wang

Subject: **Proposed**[1] meeting times

Hi, Melody:

I **received**[2] your e-mail about the series of meetings to **discuss**[3] the **budget**[4] **details**[5]. **With regard to**[6] the first meeting, yes, Friday works for me. You also suggested having a second meeting on the following Monday morning. That time is perfect. I'll talk to you then. Now, **regarding**[7] the third meeting, I'm free on the 20th, but not on the 21st. Concerning the proposed meeting after that, **I'm tied up**[8] on the morning of 25th, but the afternoon of that day works for me. For the final meeting, you **recommended**[9] the 26th. However, can we make it for another time? Sorry, that day is impossible for me.

Sincerely,
Frank
Aim Communications

中译

收件人：梅洛迪·肯特
发件人：弗兰克·王
主　题：会议的建议时间

梅洛迪：

你好！

我已收到有关讨论预算细节系列会议的电子邮件。关于第一次会议，是的，星期五可以。你也建议在接下来的星期一早上开第二次会。那个时间我没问题。那就到时再和你谈。现在关于第三次会议，我 20 号有空，但 21 号没空。而有关之后的会议，我 25 号整个上午都很忙，但那天下午我可以到会。至于最后一次会议，你建议在 26 号召开，但我们能不能改约另一个时间？抱歉，那天我不行。

标的通讯公司
弗兰克　敬上

Vocabulary and Phrases

1. **proposed** [prəˋpozd] *a.* 被提议的
   propose [prəˋpoz] *vt.* 提议，建议
   = suggest
   propose + V-ing　提议……
   例: I proposed changing the name of the company.
   （我提议更改公司名称。）

2. **receive** [rɪˋsiv] *vt.* 收到
   例: I received a phone call from your father the other day.
   （我几天前接到你父亲的电话。）

3. **discuss** [dɪˋskʌs] *vt.* 讨论
   discuss sth with sb　与某人讨论某事
   例: I'm not prepared to discuss this with you.
   （我尚未准备好要与你谈论此事。）

4. **budget** [ˋbʌdʒɪt] *n.* 预算

5. **detail** [ˋditel] *n.* 细节
   in detail　详细地
   例: This issue will be discussed in more detail in the next chapter.
   （这个问题在下一章会详细论述。）

6. **with regard to...**　关于……
   = concerning...
   = regarding...
   = about...
   例: I'm writing to you with regard to your letter of March 15.
   （我写这封信是回复你 3 月15 日的来信。）

7. **regarding** [rɪˋgardɪŋ] *prep.* 关于
   例: Call me if you have any problems regarding your work.
   （你如果还有什么工作方面的问题就给我打电话。）

8. **tie sb up**
   使某人很忙碌（常用被动语态）
   例: I'm tied up in a meeting until 5.
   （我开会开到 5 点才能脱身。）

9. **recommend** [ˌrɛkəˋmɛnd] *vt.* 建议
   = suggest
   recommend + V-ing　建议……
   = suggest + V-ing
   例: I recommend reading the book before seeing the movie.
   （我建议先看这本书，再去看这部电影。）

Business Writing Exercises

请按括号中的提示将下列句子译成英文。

1. 我星期四上午有空，但星期三下午没空。

   ____

2. 6 月 8 日我可以。那你呢？

   ____

3. 恐怕我明天一整天都极为忙碌。（be tied up）

   ____

4. 星期五下午 3 点比较适合我。（be preferable to）

   ____

5. 关于开会日期，你有什么建议吗？（With regard to...）

   ____

Unit 3

Rescheduling and Canceling Meetings 改期和取消会议

Basic Structure 基本结构

1. 点出会议将被改期（I need to postpone our meeting until...）或取消（The meeting has been canceled.）。
2. 说明原因（I'm afraid something has come up.）。
3. 告知新的会议时间（The time has been changed to...）并致歉（I apologize for the change.）。

- 如要改期或取消会议，应马上通知相关人。
- 在要求改期或取消会议时，可以将 Unfortunately 或 Sorry 等词放在句首表示遗憾之意，以避免触怒对方，使人不悦。例如：
  Unfortunately, I'm unable to make the meeting scheduled for Friday morning.（不巧的是，我无法出席预定在星期五早上的会议。）

Sentence Patterns 写作句型  06

I'm afraid something has come up. / Regretfully, my schedule has changed.
我恐怕临时有事。/ 遗憾的是，我的日程安排有所更改。

- I'm afraid something has come up, so I can't keep our appointment.
  我恐怕临时有事，所以我无法赴约。
- Regretfully, my schedule has changed this week.
  遗憾的是，我这星期的日程安排有所更改。

Unfortunately, I'm unable to make the meeting scheduled for (date / day / time). / Sorry, but I forgot about (an event / another meeting / appointment, etc.).
不巧的是，我无法于（某日期 / 某天 / 某时间）参加会议。/ 抱歉，我忘记（某事件、其他会议、约定等）。

- Unfortunately, I'm unable to make the meeting scheduled for Friday morning.
  不巧的是，我无法出席预定在星期五早上召开的会议。
- Sorry, but I forgot about a colleague's retirement dinner next Saturday.
  抱歉，我忘记下星期六有同事的退休晚宴。

I need to postpone our meeting until (date / day / time). / Can we reschedule it to (date / day / time)?
我需要将我们的会议延期至（某日期 / 某天 / 某时间）。/ 我们能改期至（某日期 / 某天 / 某时间）吗？

- I need to postpone our meeting until later this week or next week.
  我需要将我们的会议延期至本星期稍后或下星期。
- Can we reschedule it to August 8?
  我们能把它改期至 8 月 8 日吗？

The time for the meeting has been changed to (date / day / time). / Due to circumstances, the meeting (date / day / time) has been canceled / postponed.
会议的时间已更改为（某日期 / 某天 / 某时间）。/ 基于某些情况，（某日期 / 某天 / 某时间）的会议已被取消 / 延后。

- The time for the meeting has been changed to Thursday.
  会议的时间已更改为星期四。
- Due to circumstances, the meeting on Monday has been canceled.
  基于某些情况，星期一的会议已被取消。

Sorry for any inconvenience. / I apologize for the change.
不便之处，敬请见谅。/ 对于这项变更我感到抱歉。

- Sorry for any inconvenience this has caused.
  对于此事所造成的不便之处，敬请见谅。
- I apologize for the change to the schedule.
  对于日程安排的更改我感到抱歉。

Model E-mail 电子邮件范例  06

To: Raymond Xu, Assistant **Advertising**[1] Manager, Ad Planet

From: Stephanie Adamson, Advertising Manager, Ad Planet

Subject: The meeting on May 2

Hi Ray:

With regard to the advertising meeting next Wednesday, I'm afraid something has **come up**[2]. One of my **major**[3] clients is coming to town and he can only meet with me on May 2. This means that, unfortunately, I'm unable to make our meeting **scheduled**[4] for that day. Can we reschedule the advertising meeting to **either** May 3 **or**[5] May 4?
Could you also inform the others about this change? Sorry for any **inconvenience**[6] this may cause you, Ray.

Yours truly,
Stephanie

To: Dave Fang, Mary Black, Harrison Ping, Fred Willis

From: Raymond Xu

Subject: Weekly Managers' Meeting

Dear Dave, Mary, Harrison, and Fred:

Due to[7] **circumstances**[8], the meeting on May 2 has been **postponed**[9]. Regretfully, Stephanie's schedule has changed and she can't make the meeting. The new date for the meeting is May 4 at 10:30 a.m. Stephanie apologizes for the change.

Best regards,
Ray

中译

收件人：星球广告公司　广告部副经理
雷蒙德・许
发件人：星球广告公司　广告部经理
斯蒂芬妮・亚当森
主　题：5月2日的会议

雷：

你好！

关于下星期三的广告会议，我临时突然有事。我的一位重要客户要进城，他只有5月2日有空见我。也就是说，很不巧我无法参加当天的会议。我们可以重新安排广告会议至5月3日或5月4日吗？麻烦你也通知其他人此项变更。造成你的任何不便，我深感抱歉。

斯蒂芬妮　敬上

中译

收件人：戴夫・方、玛丽・布莱克、
哈里森・平、弗雷德・威利斯
发件人：雷蒙德・许
主　题：每周经理会议

亲爱的戴夫、玛丽、哈里森及弗雷德：

基于某些情况，5月2日的会议已被延期。遗憾的是，斯蒂芬妮的行程有所变动，所以她不能参加会议。新的会议时间为5月4日上午10点30分。斯蒂芬妮为此改期感到抱歉。

诚挚的祝福

雷

Vocabulary and Phrases

1. **advertising** [ˋædvə͵taɪzɪŋ] *n.* 广告(业)(集合名词,不可数)
   advertisement [ˋædvə͵taɪzmənt] *n.* 广告(可数,缩写为 ad)
   例: A good advertising campaign will increase our sales.
   (好的广告活动会增加我们的销售量。)
   Tom put an ad in the local paper to sell his old car.
   (汤姆在当地报纸刊登广告出售他的老爷车。)

2. **come up** 发生;出现
   例: We'll let you know if any vacancies come up.
   (如有任何职缺我们会告知你。)
   * vacancy [ˋvekənsɪ] *n.* 工作职缺

3. **major** [ˋmedʒə] *a.* 重要的
   play a major role in... 在……起重要作用
   例: Peter plays a major role in our company.
   (彼得在本公司起着重要的作用。)

4. **schedule** [ˋskɛdʒʊl] *vt.* 为……安排时间
   be scheduled for + 时间 ……安排在某时间
   be scheduled to V 预定从事……
   例: The director meeting is scheduled for Thursday afternoon.
   (主管会议安排在星期四下午。)
   Our client is scheduled to arrive in Beijing at 5 o'clock.
   (我们的客户定于 5 点抵达北京。)

5. **either A or B** 不是 A 就是 B(连接两个主语时,动词根据后者作变化)
   例: Either you or <u>he</u> <u>is</u> wrong.
   (不是你就是他错。)
   I'm going to buy either a camera or a cell phone with the money.
   (我要用这笔钱去买相机或是手机。)

6. **inconvenience** [͵ɪnkənˋvinjəns] *n.* 不便

7. **due to...** 由于……
   = because of...
   = owing to...
   = as a result of...
   例: Due to the rain, the game was canceled.
   (因为下雨,那场比赛取消了。)

8. **circumstance** [ˋsɜkəm͵stæns] *n.* 情况(常用复数)
   under no circumstances 绝不……(常置于句首,之后采用问句式倒装句型)
   = by no means
   例: <u>Under no circumstances</u> <u>should you</u> quit your job now.
   = <u>By no means</u> <u>should you</u> quit your job now.
   (目前你绝对不可以辞职。)

9. **postpone** [poˋspon] *vt.* 延期
   postpone + N/V-ing 延后……
   例: They decided to postpone building the new hospital until next year.
   (他们决定将新医院的修建延期到明年。)

Business Writing Exercises

请按括号中的提示将下列句子译成英文。

1. 恐怕我们必须立即采取行动。(take immediate action)
   ____
2. 不幸的是,我抵达机场时飞机已起飞。(take off)
   ____
3. 你可否将这次会议改期至下星期一?
   ____
4. 由于情况非我们所能控制,我们无法赶上最后期限。(meet the deadline)
   ____
5. 延误造成的不便,敬请见谅。
   ____

Unit 4

Outlining Agendas　议程大纲

Basic Structure　基本结构

1. 表明议程大纲已附在邮件中，请收件人与会前先行阅读。（Please refer to the proposed agenda in the attachment.）
2. 表明收件人若对议程大纲有新增建议，请于特定时间内回信告知。（If you have any additions to it, please inform me by...）
3. 简单说明议程大纲。（The first item on the agenda is...）
4. 提醒此次会议大约持续时间。（The meeting is scheduled to last...）

- 此类邮件目的在于让与会者了解会议讨论的主题，并可根据此信息做必要的准备，因此必须在召开会议前将其发出。而询问对方对议程的意见可避免遗漏某些重要事项。
- 在必要时，列出各事项的开始时间供与会者参考；假如某特定事项是由某人负责做简报，可以把他的名字以括号标注在细目旁边。

Sentence Patterns 写作句型 07

We have a heavy agenda. / The agenda is quite full.
我们这次的议程相当紧凑。

- We have a heavy agenda, so please don't be late.
  我们这次的议程很紧凑，所以请别迟到。
- The agenda is quite full for the meeting on Monday.
  星期一会议的议程相当紧凑。

Attached is a copy of the tentative agenda. / Please refer to the proposed agenda in the attachment.
兹附上一份暂定议程。 / 请参照附件中拟议的议程。

- Attached is a copy of the tentative agenda for the managers' meeting.
  兹附上一份本次经理会议的暂定议程。
  * tentative [ˋtɛntətɪv] *a.* 暂定的
- Please refer to the proposed agenda in the attachment to this e-mail.
  请参照本电子邮件附件中拟议的议程。

The first item on the agenda is... / The first issue we will deal with is...
议程的第一项是…… / 我们首要处理的议题是……

- The first item on the agenda is a report on last month's sales figures.
  议程的第一项是上个月的销售数字报告。
- The first issue we will deal with is the proposed budget.
  我们首先要处理的议题是预算提案。

Other matters to be discussed include... / Also on the table for discussion is / are...
其他将要讨论的事项包括…… / 此外还需提交讨论的是……

- Other matters to be discussed include travel expenses and overtime pay.
  其他将要讨论的事项包括差旅费及加班费。
- Also on the table for discussion are the problems with the new factory.
  此外还需提交讨论的是新工厂的问题。

The meeting is scheduled to last... / The meeting should take (time)
本会议预计将持续…… / 本会议应该会耗时……

- The meeting is scheduled to last two hours.
  本会议预计将持续两小时。
- The meeting should take about an hour and a half at the most.
  本会议应该会耗时最多约一个半小时。
  * at (the) most 至多；不超过

Model E-mail 电子邮件范例 07

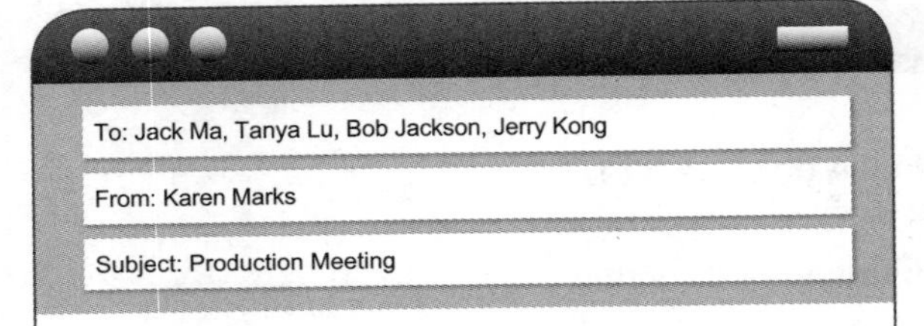
To: Jack Ma, Tanya Lu, Bob Jackson, Jerry Kong

From: Karen Marks

Subject: Production Meeting

Hi Jack, Tanya, Bob and Jerry:

Regarding Monday's production meeting, we have a **heavy**[1] **agenda**[2], so please arrive on time. Attached to my e-mail is a copy of the **tentative**[3] agenda. Please **refer to**[4] the proposed agenda before the meeting. If you have any questions about the agenda, or **additions**[5] to it, please **inform**[6] me by this Thursday **at the latest**[7]. As you will see, the first **item**[8] on the agenda is cutting production costs. Also on the table for discussion is the budget for new **machinery**[9] for next year. Other matters to be discussed include ways to improve efficiency and **quality control**[10]. The meeting should take about three or four hours.

Regards,
Karen

中译

收件人：杰克·马、坦尼娅·卢、鲍勃·杰克逊、杰瑞·孔
发件人：凯伦·马克思
主　题：生产会议

杰克、坦尼娅、鲍勃、杰瑞，诸位好：

事关星期一的生产会议，由于议程紧凑，请诸位准时前来开会。这份电子邮件另附暂定议程附件一份。请于开会前参照拟议的议程。诸位若对议程有任何疑问，或添加事项，请最晚于本星期四前通知我。如诸位所见，议程第一项是削减生产成本。此外还需提交讨论的是下年度新机器设备预算案。其他讨论事项则包括提高效率及质管的方法。本次会议需时 3 至 4 小时。

诚挚的祝福

凯伦

Attachment

Hi everyone:

Here is the tentative agenda for Monday's production meeting.

8:00 a.m. Introductory comments by Stephen Hu, senior production manager

8:10 a.m. Discussion on proposals to cut production costs

10:00 a.m. Budget for new machinery (led by Bob Jackson)

10:30 a.m. Efficiency review (Tanya Lu)

11:15 a.m. Report on quality control (Karen Marks)

12:00 p.m. lunch (provided)

If you have any questions about the agenda, or additions to it, please inform me by this Thursday at the latest.

-Karen

中译

【附件】

诸位好：

以下是星期一生产会议的暂定议程。

上午 8 点
由高级生产经理史蒂文·胡发表引言

上午 8 点 10 分
针对削减生产成本提案进行讨论

上午 10 点
新机器设备的预算（由鲍勃·杰克逊带领）

上午 10 点 30 分
效率检讨（坦尼娅·卢）

上午 11 点 15 分
质管报告（凯伦·马克思）

中午 12 点
午餐（公司提供）

诸位若对议程有任何疑问，或添加事项，请最晚于本星期四前通知我。

凯伦

Vocabulary and Phrases

1. **heavy** [ˈhɛvɪ] *a.* 繁重的
   例: I have a heavy schedule for the next few days.
   （我今后几天日程安排紧得要命。）
2. **agenda** [əˈdʒɛndə] *n.* 议程；日常事项
   be high on the agenda
   优先待办事项，首要事项
   例: There are several important items on the agenda.
   （议程上有几件重要的事项。）
   Quality is high on the agenda in our company.
   （本公司把质量放在日常工作的首位。）
3. **tentative** [ˈtɛntətɪv] *a.* 暂定的
   例: We made a tentative arrangement to meet again next Monday.
   （我们暂定下星期一再碰面。）
4. **refer to sth** 查阅/参照某物
   例: You may refer to your notes during the listening test.
   （你在听力考试中可以参阅笔记。）
5. **addition** [əˈdɪʃən] *n.* 添加事项（与介词to连用）
   例: Welcoming a new addition to the family is very exciting.
   （欢迎家中新成员的到来令人相当兴奋。）
6. **inform** [ɪnˈfɔrm] *vt.* 通知
   = notify [ˈnotəˌfaɪ]
   inform sb of sth 通知某人某事
   = notify sb of sth
   例: Please inform me of your final decision.
   （请通知我你最终的决定。）
7. **at the latest** 最晚
   at the earliest 最早
   例: Application forms should be turned in by May 1 at the latest.
   （申请表最晚于5月1日以前提交。）
8. **item** [ˈaɪtəm] *n.* 事项
9. **machinery** [məˈʃinərɪ] *n.* 机器（集合名词，不可数）
   machine [məˈʃin] *n.* 机器（可数）
   a piece of machinery 一台机器
   = a machine
   some machinery 一些机器
   = some machines
10. **quality control** 质量管理（简称QC）

Business Writing Exercises

请按括号中的提示将下列句子译成英文。

1. 时间快不够用了，所以我们移到议程的最后一项吧。(Time is running short / out, so let's...)

   ____

2. 兹附上本人简历附件一份。

   ____

3. 我们首先要处理的议题是我们的一条新生产线。(...one of our new production lines)

   ____

4. 此外还需提交讨论的是我们是否应该在海外进行更多的投资。(...make more investments abroad)

   ____

5. 本次会议预计将持续最多3小时。

   ____

Unit 1

Making Travel Arrangements 安排行程

Basic Structure 基本结构

1. 表明即将前往何地出差。（I am going on a business trip to...）
2. 说明时间及行程。
3. 说明可会面的地点及会面形式（正式开会或聚餐）。（Would it be possible to visit you at your office, or enjoy lunch together?）

- 这类电子邮件目的是向对方说明自己的行程，并询问对方能否见面，因此内容可包括自己的行程和欲讨论的事项，以供对方安排后续事宜。
- 在结束正文时，会习惯性地礼貌问候，而最常见的结尾敬词是 Yours faithfully（敬启，用在称呼为 Dear Sir / Sirs 或 Dear Madam 时），Yours sincerely（敬上，用在知道收件人是谁时，如：Dear Sam, Dear Mr. Bagnell）及 Yours truly（此致，用在不太熟悉的人时）。

Sentence Patterns 写作句型 08

I'll be in (location) for (time period). / I'm planning a trip to (location) in (month) /from (date) to (date).

我会在（某地点）待上（一段时间）。/ 我正计划在（某月）/ 从（某日期）到（某日期）前往（某地）旅行。

- I'll be in Toronto for a week in October.
  我 10 月份会在多伦多待一个星期。
- I'm planning a trip to London from March 4 to March 12.
  我正计划 3 月 4 日到 12 日去一趟伦敦。

I'll be arriving in (location) on (date). / My flight gets into (location) at / on (time /date).

我将在（某日期）抵达（某地）。/ 我的班机将在（某时间 / 某日期）到（某地）。

- I'll be arriving in Tokyo on Tuesday, January 21.
  我将在 1 月 21 日星期二抵达东京。
- My flight gets into Seoul at 11 a.m. on Sept. 1.
  我的班机将在 9 月 1 日上午 11 点抵达首尔。

I would really like to meet with you about... / I'd love to be able to talk with you about...

我很想与你见面洽谈（有关）…… / 我很乐意能与你洽谈（有关）……

- I would really like to meet with you about the proposed deal.
  我很想与你见面洽谈那项拟议中的交易。
- I'd love to be able to talk with you about our new products.
  我很乐意能与你洽谈本公司的新产品。

Perhaps we could... / Would it be possible to...?

或许我们可以…… / 是否可以……?

- Perhaps we could get together while I'm in Paris.
  我在巴黎时，或许我们可以聚一聚。
- Would it be possible to meet for lunch on the 6th of October?
  是否可以在 10 月 6 日共进午餐？

Hopefully, I can see you... / I hope our schedules work out so that...

希望能在……见到你。/ 希望我们的日程安排相互配合，以便能……

- Hopefully, I can see you next Wednesday.
  希望能在下星期三见到你。
- I hope our schedules work out so that we can meet in August.
  希望我们的日程安排相互配合，以便能在 8 月会面。

Model E-mail 电子邮件范例 08

To: Fanny Carlton

From: Joseph Wang

Subject: Trip to Chicago

Dear Fanny:

How are you? I'll be in the United States for a **couple**[1] of weeks in May. In fact, I'll be traveling to the Chicago area on May 12, and I'm planning to be there for three or four days. If you have time, I would really like to meet with you about our new **line**[2] of tools. Perhaps we could meet while I am in Chicago. Would it be possible to visit you at your office, or enjoy lunch together?
I hope our **schedules**[3] **work out**[4] so that we can hook up. I look forward to talking with you again.

Joseph

中译

收件人：范妮·卡尔顿
发件人：约瑟夫·王
主　题：芝加哥之行

亲爱的范妮：

你好吗？ 5 月份我会在美国待几个星期。事实上，我会在 5 月 12 日去一趟芝加哥，计划在那里待上 3 到 4 天。如果你有空，我很想与你见面洽谈本公司新系列的工具。或许我在芝加哥时，我们可以见个面。是否能前往你的办公室拜访，或共进午餐？

希望我们的日程安排相互配合，以便能够相聚。期待再次与你谈话。

约瑟夫

Vocabulary and Phrases

1. **couple** [ˋkʌpl̩] *n.* 几个（人、物）
   a couple of + 复数名词
   几个……（等于 a few...）
   例: I'll be with you in a minute. There are a couple of things I have to do first.
   （我一会儿就会去你那儿。我有几件事得先处理一下。）
2. **line** [laɪn] *n.* 生产线（= production line）；产品系列（= product line）
   例: Cars are checked as they come off the production line.
   （汽车下了生产线立即进行校验。）
3. **schedule** [ˋskɛdʒʊl] *n.* 日程安排；进度表
   on schedule 照进度
   ahead of schedule 进度超前
   fall behind schedule 进度落后
   a tight / heavy schedule 日程紧凑
   例: Thanks to your help, I managed to finish the book on schedule / ahead of schedule.
   （多亏你的帮忙，我如期／提早写完这本书。）
   The project fell behind schedule due to the manager's inability.
   （这个项目因为经理的无能而进度落后。）
   We can't take on any extra work. We're already working on a tight schedule.
   （我们无法承担任何额外的工作。我们的进度表已经排得很满了。）
   * take on sth 承担（责任等）
4. **work out** 成功地发展
   例: Things have worked out quite well for us.
   （事情的进展对我们很有利。）

Business Writing Exercises

请按括号中的提示将下列句子译成英文。

1. 我正计划从 5 月 10 日至 17 日前往纽约作一趟商务旅行。

   ____

2. 我的班机大约于早上 10 点抵达，如果没有延误的话。

   ____

3. 我实在很想与你会面洽谈我们的合作计划。（our cooperation plans）

   ____

4. 你可以开车到机场接我吗？（pick me up）

   ____

5. 希望我们以后能更常保持联系。（keep in touch / contact）

   ____

Unit 2

Dealing with Hotel Reservations
处理饭店预订

Basic Structure 基本结构

1. 确认饭店住宿时间及房型。(You are scheduled to stay here for...)
2. 确认入住事宜。(The check-in time is...)
3. 告知前往饭店方式。(You can catch a shuttle bus to the hotel.)

- 这类电子邮件的目的是向对方说明所安排的饭店信息，内容应包括住房日期、天数及确切的停留时间。例如：
  You are scheduled to stay there for three nights, from March 9 to March 12. (你预计要在那儿停留3个晚上，从3月9日到3月12日)；此外，在信中也可注明登记入住和退房的时间、饮食及周边交通情况。
- 须提供饭店提供的预订确认号码。

Sentence Patterns 写作句型 09

I've booked a room at (hotel name) for (period of time). / I've made reservations at (hotel name) for (period of time).
我已在（某饭店）订了房间，住（某段时间）。

- I've booked a room at the Harrison Springs Hotel for four nights.
  我已在哈里森温泉饭店预订住房 4 晚。
- I've made reservations at the Crystal Lake Lodge for six nights.
  我已在水晶湖山庄预订住宿 6 晚。
  * lodge [lɑdʒ] *n.* 原指“乡间小屋”，此处做专有名词，指“度假屋”。

Your reservation is confirmed at (hotel name). / Your confirmation number is (number).
您在（某饭店）的预订已获得确认。/ 您的确认号码是……。

- Your reservation is confirmed at the Redstone Manor.
  您在红石庄园的预订已获得确认。
  * manor [ˋmænɚ] *n.* 大庄园
- Your confirmation number is ES329BL.
  您的确认号码为 ES329BL。

The check-in time is (time). / The check-out time is (time).
办理入住时间为（某时间）。/ 办理退房时间为（某时间）。

- The check-in time is 3 p.m. / The check-out time is 12 p.m.
  办理入住时间是下午 3 点。/办理退房时间是中午 12 点。

There is a complimentary breakfast. / Breakfast is included in the price of the room.
早餐免费提供。/ 早餐已包含在房费内。

- There is a complimentary breakfast served from 7 a.m. to 10:30 a.m.
  早上 7 点至 10 点半提供免费早餐。
  * complimentary [ˌkɑmpləˋmɛntərɪ] *a.* 免费的
- Breakfast is included in the price of the room at the Grayson Inn.
  格雷森饭店的房费内包含早餐。

You can catch a shuttle bus to the hotel. / There are taxi stands (outside / near...).
您可搭乘接送巴士至该饭店。/（某位置）有出租车招呼站。

- You can catch a shuttle bus to the hotel on the ground level.
  您可至一楼搭乘接送巴士至本饭店。
  * ground floor 底楼，一楼（建筑物与地面相平的一层）
- There are taxi stands outside the terminal, near the bus stop.
  航站楼外的公交车站附近有出租车招呼站。

Model E-mail 电子邮件范例 09

To: Mr. Bosworth

From: Tina Zhang

Subject: Business trip to Spain

Dear Mr. Bosworth:

Regarding your **upcoming**[1] business trip to Barcelona, I've **booked**[2] a room at the Matador **Inn**[3]. You are scheduled to stay there for six nights. Your reservation is **confirmed**[4], and your confirmation number is YXR-30891. According to the hotel's **website**[5], the check-in time is 2 p.m. and the check-out time is noon. Breakfast is included in the price of the room. I believe it is at a **buffet**[6] restaurant on the second floor.

After you **land**[7] in Barcelona and go through **customs**[8] and **immigration**[9], you can catch a shuttle bus to the hotel. The shuttle bus stop is located on the lower level of the airport. There are also taxi stands outside the **terminal**[10] if you prefer to take a **cab**[11].

Have a good trip to Spain!

Sincerely,
Tina

中译

收件人：波茨沃斯先生

发件人：蒂娜·张

主　题：前往西班牙出差

亲爱的波茨沃斯先生：

关于您将前往巴塞罗那出差一事，我已在斗牛士饭店为您预订房间。您预计在那儿住6晚。您的订房已获得确认，确认号码为 YXR-30891。根据饭店网站所述，办理入住手续时间是下午 2 点，办理退房手续时间是中午。早餐包含在房价内。我想早餐是在 2 楼的自助餐厅内。

班机降落至巴塞罗那后，您先通过海关和移民局检查站，接着可搭乘接送巴士至该饭店。接送巴士站位于机场地下层。如果您比较喜欢乘出租车，航站楼外也有出租车招呼站。

敬祝西班牙之旅旅途愉快！

蒂娜 敬上

Vocabulary and Phrases

1. **upcoming** [ˈʌpˌkʌmɪŋ] *a.* 即将来临的
   例: Tickets are selling well for the singer's upcoming concert.
   （该歌手即将举办的演唱会票卖得很好。）
2. **book** [bʊk] *vt.* 预订
   = reserve [rɪˈzɜːv]
   reservation [ˌrɛzəˈveʃən] *n.* 预订
   book sth
   预订（票、饭店房间、桌位等）
   = reserve sth
   = make a reservation for sth
   例: I'd like to make a reservation for a table for four.
   （我想订个 4 人的桌位。）
3. **inn** [ɪn] *n.* 客栈，小旅馆（尤用于旅馆或饭店专有名词中）
4. **confirm** [kənˈfɜm] *vt.* 确认
   confirm a booking / a reservation
   确认预订
   例: I'm calling to confirm a reservation for a double room for July 31.
   （我来电是要确认 7 月 31 日一间双人房的预订。）
5. **website** [ˈwɛbˌsaɪt] *n.* 网站
   例: For more information, you can log on to our website.
   （要取得更多信息，您可登录我们的网站。）
6. **buffet** [bəˈfe] *n.* 自助餐
7. **land** [lænd] *vi. & vt.* （使）着陆
   例: We shall be landing shortly. Please fasten your seatbelt.
   （我们很快就要着陆，请您系好安全带。）
   The pilot landed the plane safely.
   （飞行员驾驶飞机安全着陆。）
8. **customs** [ˈkʌstəmz] *n.* 海关（常用复数）
   go through customs　通过海关
9. **immigration** [ˌɪməˈgreʃən] *n.*
   （机场的）移民检查局
10. **terminal** [ˈtɜmənl̩] *n.* 航站楼
11. **cab** [kæb] *n.* 出租车
    cabby [ˈkæbɪ] *n.* 出租车司机

Business Writing Exercises

请按括号中的提示将下列句子译成英文。

1. 我以你的名义已在那家饭店订了一个房间。（... in your name）

   ____

2. 我来电告知您的预订机位已获得确认。（seat reservation）

   ____

3. 大多数饭店的退房时间是中午。

   ____

4. 早餐包含在房费里吗?

   ____

5. 如果你不想乘出租车，你可以坐接送巴士到我们饭店。

   ____

Unit 3

Confirming an Itinerary 确认行程

Basic Structure 基本结构

1. 确认飞机航班号及起飞、抵达时间。
2. 若非直飞航班（a direct flight），则须说明中途停留地（stopover）及停留时间。
3. 确认观光行程及接送方式。

- 这类电子邮件的目的是向对方说明所安排的航班行程，内容应包括航空公司名称、航班号、起飞及抵达当地的时间。例如：
  You are booked on North American Airlines, Flight NA812. The plane departs at 3:30 p.m. local time and arrives in New York at 6:15 p.m., New York time.（您预订的是北美航空公司，NA812 号航班。该航班预计于当地时间下午 3 点 30 分起飞，于纽约时间下午 6 点 15 分抵达纽约。）
- 如果所安排航班和对方要求有所差异，应诚实地向对方说明。

Sentence Patterns 写作句型 10

Here are the trip details. / Here is the itinerary for the trip.
以下是本次旅程的细节事项。 / 以下是本次旅行的行程。

- Here are the flight details for the business trip.
  以下是本次出差的航班细节。
- Here is the itinerary for the trip to Dallas next month.
  以下为下个月前往达拉斯之行的行程。

The (transportation) departs at (time) / The (transportation) arrives in (location) at (time).
（交通工具）预计于（某时间）起程。 / （交通工具）预计于（某时间）抵达（某地）。

- The plane departs at 11 a.m. from Terminal 1.
  该航班预计于上午 11 点从第一航站楼起飞。
- The plane arrives in Prague at 6:45 p.m.
  该航班预计于下午 6 点 45 分抵达布拉格。

The flight from (location) to (location) is a direct flight. / There is a (time) stopover / layover in (location).
该航班从（某地）直飞（某地）。 / 中途将在（某地）停留（某时间）。

- The flight from Los Angeles to Hong Kong is a direct flight.
  该航班从洛杉矶直飞香港。
- There is a four-hour stopover in Tokyo.
  中途将在东京停留 4 小时。

Please be (location) at (time). / You will be picked up (location) at (time).
请于（某时间）到达（某地）。 / （某时间）将有人去（某地）接你。

- Please be in the hotel lobby at or before 10 a.m.
  请于上午 10 点整或提早至饭店大厅等候。
- You will be picked up from your hotel at 9 a.m.
  上午 9 点将有人去饭店接您。

The tour starts at (time). / The sightseeing trip ends at (time).
旅程从（某时间）开始。 / 观光于（某时间）结束。

- The half-day tour of the city starts at 1 p.m.
  市区半日游从下午 1 点开始。
- The sightseeing trip of San Francisco ends at 4:30 p.m.
  旧金山市区观光于下午 4 点 30 分结束。

Model E-mail 电子邮件范例 10

To: Ms. Luis

From: Daphne Han, Terrific Trips Travel Agency

Subject: Final **itinerary**[1]

Dear Ms. Luis:

Further to our conversation this morning, here is the itinerary for your trip. You are booked on North American Airlines, Flight NA812. The plane **departs**[2] at 3:30 p.m. local time and arrives in New York at 6:15 p.m., New York time. Unfortunately, I couldn't **arrange**[3] **a direct flight**[4] for you, so there is a two-hour **stopover**[5] in Seattle. According to your **request**[6], I have also booked a full-day city tour on April 16, including stops at the Statue of Liberty and Times Square. You will be **picked up**[7] outside the main **entrance**[8] of your hotel at 8:30 a.m. The **sightseeing**[9] trip ends at 6 p.m. and you will get back to your hotel before 7 p.m.

Bon voyage!
Daphne Han

中译

收件人：路易丝女士
发件人：顶呱呱旅行社　达芙妮·韩
主　题：最终行程安排

亲爱的路易丝女士：

针对我们今天上午的谈话，以下是您本次旅行的行程。您预订的是北美航空公司，NA812 号航班。该航班预计于当地时间下午 3 点 30 分起飞，于纽约时间下午 6 点 15 分抵达纽约。遗憾的是，我没能为您安排到直飞航班，因此您中途会在西雅图停留两小时。根据您的要求，我也订了 4 月 16 日的城市一日游，包括在自由女神像及时代广场等地短暂停留。上午 8 点 30 分将有人在饭店大门外接您。观光行程将于下午 6 点结束，您将于晚上 7 点前返抵饭店。

祝一路平安！

达芙妮·韩

Vocabulary and Phrases

1. **itinerary** [aɪˋtɪnəˌrɛrɪ] *n.* 旅程安排
2. **depart** [dɪˋpɑrt] *vi.* 出发
   depart for + 地点　出发至某地
   = leave for + 地点
   = set out for + 地点
   = set off for + 地点
   例: Our manager departed for New York the other day.

（我们经理前几天出发到纽约去了。）

3. **arrange** [əˈrendʒ] *vt. & vi.* 安排；筹备
   arrange a meeting / an appointment
   安排会议 / 约见
   arrange for sb / sth to V
   安排某人 / 某物……
   例: You can contact our head office and arrange a meeting with our marketing director.
   （你可以联络本公司总部，与市场总监安排一次会议。）
   I've arranged for a car to pick us up at the airport.
   （我已安排一辆车到机场接我们。）
4. **a direct flight** 直飞航班
   = a non-stop flight
5. **stopover** [ˈstɑpˌovɚ] *n.* 中途停留
   = layover [ˈleˌovɚ]
   例: We had a stopover in Japan on the way to the US.
   （我们去美国途中在日本中途停留。）
6. **request** [rɪˈkwɛst] *vt. & n.* 请求
   request sb to V 请求某人……
   request that + 主语 + (should) V
   请求……
   make a request for sth
   针对某事提出请求
   at the request of sb
   按照某人的请求
   例: We were requested to assemble in the lobby.
   （我们被要求到大厅集合。）
   Tom requested that no one (should) be told of his final decision.
   （汤姆请求不要告诉任何人他最终的决定。）
   The war-torn country made a request for international aid.
   （那个饱受战争摧残的国家请求国际援助。）
   Henry carried out marketing research at the request of his manager.
   （亨利按照经理的要求进行了市场调查。）
7. **pick sb up** （开车）接某人……
   例: Jason will pick me up at 10:30 a.m.
   （杰森上午10点30分会开车来接我。）
8. **entrance** [ˈɛntrəns] *n.* 入口
   the main entrance 正门
   例: I'll meet you at the main entrance.
   （我会在正门等你。）
9. **sightseeing** [ˈsaɪtˌsiɪŋ] *n.* 观光
   go sightseeing 去观光
   例: Since we have to leave tomorrow, there's no time for us to go sightseeing.
   （因为我们明天就要离开了，所以没时间去观光。）

Business Writing Exercises

请按括号中的提示将下列句子译成英文。

1. 以下是你们环岛旅行的行程。（around the island）
   ____
2. 该航班预计于下午 5 点从第二航站楼起飞。
   ____
3. 中途将在台北停留两小时。
   ____
4. 按照您的要求，将有人去机场接您。（According to your request, you...）
   ____
5. 如果您喜欢的话，我们可以为您安排直飞航班。（If you like, we could...）
   ____

Unit 1

Asking for Recommendations 寻求建议

Basic Structure 基本结构

1. 点出来信旨在寻求建议。(Do you have any recommendations regarding...?)
2. 清楚说明需要建议的事项。(Can you recommend a / any...?)
3. 表明愿意详谈。(We really should get together sometime to have a further discussion.)

- 在请求建议时的语气要视你和收件人的关系而定，例如请求上级建议时会比请下属建议更恭敬。
- 在询问完对方的建议后，可以用“Thanking you in advance for your help in this matter.”(先在此感谢您在这件事上对我的帮助。)来作为结束语。

Sentence Patterns 写作句型 11

Do you know a / any (...)? / Are you familiar with a / any (...)?
你认识某个 / 什么（……）吗？ / 你和某个 / 什么（……）熟吗？

- Do you know any decent plumbers?
  你认识什么还不错的水暖工吗？
  * decent [ˋdisn̩t] *a.* 像样的
  plumber [ˋplʌmɚ] *n.* 水暖工（b 不发音）
- Are you familiar with any good mechanics?
  你熟识什么不错的维修工吗？

Can you recommend a / any (...)? / Are there any (...) you would recommend?
你能推荐一个 / 什么（……）吗？ / 有什么（……）是你会推荐的吗？

- Can you recommend a good dentist?
  你能推荐一位好牙医吗？
- Are there any consultants you would recommend?
  有什么顾问是你可以推荐的吗？

Do you have any recommendations regarding (...)? / What would you suggest regarding (...)?
关于（……）你有什么建议吗？ / 关于（……）你会提出什么建议？

- Do you have any recommendations regarding doctors?
  关于医生你有什么建议吗？
- What would you suggest regarding interior decorators?
  关于室内设计师你会提出什么建议？
  * interior [ɪnˋtɪrɪɚ] *a.* 室内的

May I ask your opinion of (...)? / What do you think of (...)?
我能询问你对（……）的意见吗？ / 你认为（……）如何？

- May I ask your opinion of that company?
  我能询问你对那家公司的意见吗？
- What do you think of this automotive supplier?
  你认为这家汽车供应商如何？

If you were (V-ing), what would you (V)?
如果你现在……，你会……？

- If you were buying a car, what would you buy?
  你如果现在买车，你会买什么车？

Model E-mail 电子邮件范例 11

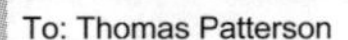

To: Thomas Patterson

From: Melody Zhang

Subject: Do you know any good suppliers?

Dear Thomas:

It's been such a long time since I last talked to you. How are things? I've been **appointed**[1] manager for an air-conditioning company. I need to find a better supplier for our products. Are you **familiar**[2] with any good suppliers?

Also, we need to hire a new **foreman**[3] for our production line. Can you recommend an **experienced**[4] foreman?

We really should get together for lunch sometime, Thomas. My **treat**[5]!

Best regards,
Melody

To: Tiffany Peng

From: Mike Richardson

Subject: Do you have any recommendations regarding carpenters?

Hi Tiffany:

I'm writing to ask for your **advice**[6]. We are **renovating**[7] the office this month. I know you did some **repairs**[8] to your office last year. What would you suggest regarding carpenters? I remember that you used a company named All-Sorts Construction. May I ask your opinion of that company? If you were renovating your office again, what would you do? Which carpenter would you choose? Would you go with All-Sorts again, or a different company?

Thanking you **in advance**[9] for your help in this matter.

Yours truly,
Mike

中译

收件人：托马斯·帕特森
发件人：梅洛迪·张
主　题：你认识什么好的供应商吗？

亲爱的托马斯：

自从我上次和你谈话后已过了很长一段时间。最近还好吗？我已被任命担任某空调公司的经理。我必须为本公司的产品找到更好的供应商。你熟识什么好的供应商吗？

此外，我们要为生产线雇用一位新工长。你能推荐一位有经验的工长吗？

托马斯，我们哪天真该找个时间共进午餐。我请客！

诚挚的祝福

梅洛迪

中译

收件人：蒂芙尼·彭
发件人：迈克·理查德森
主　题：关于木工你有什么推荐人选吗？

你好，蒂芙尼：

我来信是为了询问你的建议。我们本月正在装修办公室。我知道你去年对办公室进行了整修。关于木工你会有什么建议？我记得你雇用了一家名叫“全能”的建筑公司。我能询问你对该公司的意见吗？如果你再次整修办公室，你会怎么做？你会选哪家木工？你还是会选全能建筑公司吗？或是别家？

先在此感谢你在这件事上对我的帮助。

迈克　敬上

Vocabulary and Phrases

1. **appoint** [ə'pɔɪnt] *vt.* 任命，指派
   appoint sb (as) + 职务
   任命某人担任某职务（as 在此须省略）
   例: Mr. Miller was appointed (as) sales director.
   （米勒先生被任命为销售主管。）
2. **familiar** [fə'mɪljɚ] *a.* 熟悉的
   be familiar with... 熟悉……
   例: I'm not familiar with the computer software you use.
   （我不熟悉你们使用的电脑软件。）
3. **foreman** ['fɔrmən] *n.* 工长；领班；工头
4. **experienced** [ɪk'spɪrɪənst] *a.* 有经验的
   be experienced in... 对……有经验
   例: Tom is very experienced in marketing.
   （汤姆对营销很有经验。）
5. **treat** [trit] *n. & vt.* 请客
   例: Let's go out for lunch. It's my treat.
   （咱们出去吃午餐吧。我请客。）
   I treated Jack to lunch.
   （我请杰克吃午餐。）
6. **advice** [əd'vaɪs] *n.* 建议（不可数）
   take / follow sb's advice
   听从／遵照某人的建议
   例: Mike gave me a useful piece of advice on which computer to buy.
   （关于该买哪台电脑，迈克给我提供了一个有用的建议。）
   Take my advice—don't get married!
   （听我的劝告——别结婚！）
7. **renovate** ['rɛnə,vet] *vt.* 翻修
   renovation [,rɛnə'veʃən] *n.* 翻修
   例: The library is now under renovation.
   （那间图书馆目前正在翻修。）
8. **repair** [rɪ'pɛr] *vt.* 修理（= fix）& *n.* 修理
   beyond repair 无法修复
   例: The car is so old that it isn't worth repairing.
   （这辆车太旧了，所以不值得维修了。）
   The car was damaged beyond repair.
   （这辆车受损严重，无法修复。）
9. **in advance** 事先
   例: Did you reserve train tickets in advance?
   （你事先订火车票了吗？）

Business Writing Exercises

请按括号中的提示将下列句子译成英文。

1. 自从我们上次会面以来已经过了很久了。（It's been a long time...）
   ____
2. 你能推荐一家这附近的不错的日本料理餐厅吗？（Can you recommend...）
   ____
3. 我们真应该找个时间聚聚吃个午餐或喝杯咖啡。（We really should get together....）
   ____
4. 如果你要买个包，你会选哪个品牌？（If you were buying...）
   ____
5. 关于更进一步的阅读，你有何建议？（What would you suggest...）
   ____

Unit 2

Making Recommendations 提出建议

Basic Structure 基本结构

1. 根据对方疑问提供建议。（I suggest (that) you (should)...）
2. 希望回复的内容有所帮助。（I hope my advice is helpful to you.）

- 在提出建议时，可以使用 could、should、would、may、might 等助动词。例如：If I were you, I would invest in technology stocks.（如果我是你，我就会投资科技股。）
- 提出建议时，可多使用以下句型：
  If I were you, I would...
  I recommend you...
  I really think the best thing to do is to...
  I think you ought to...
- 在结论中你可以用"I hope my advice is helpful to you."（希望我的建议对你有帮助。）作结束语。

Sentence Patterns 写作句型 12

I suggest / recommend (that) sb (should) V
我建议某人……

- I suggest you (should) purchase a new car.
  我建议你买辆新车。
- I recommend Ben (should) start his own company.
  我建议本自己出来开公司。

Maybe you should V... / I think you ought to V...
或许你应该…… / 我想你应该……

- Maybe you should look for a new job.
  或许你应该找份新工作。
- I think you ought to choose the more experienced job candidate.
  我想你应该选择那位较有经验的求职者。

Why don't you V...? / What about + V-ing ...?
你为何不……？ / ……如何？

- Why don't you switch to Apple computers?
  你为何不换苹果电脑？
  * switch [swɪtʃ] *vi.* 转换
- What about taking a business management course?
  选一门商业管理课程如何？

Sth is famous for its... / Sth has an excellent reputation.
某物因……而出名 / 某物有绝佳的声誉。

- The Tropicana Hotel in Las Vegas is famous for its casino.
  拉斯维加斯热带饭店的赌场非常有名。
- The Peterson and Zhu law firm has an excellent reputation.
  彼得森与朱氏律师事务所有绝佳的声誉。

If I were you, I would V... / The best thing to do is to V
如果我是你，我就会…… / 最好的办法就是……

- If I were you, I would invest in technology stocks.
  如果我是你，我就会投资科技股。
- The best thing to do is to talk to a lawyer.
  最好的办法就是找个律师谈谈。

Model E-mail 电子邮件范例 12

To: Teresa Guo

From: Donald Jackson

Subject: Recommendations on insurance and investments

Dear Teresa:

Regarding your recent e-mail asking me about insurance and financial **investments**[1], most importantly, I recommend you purchase personal life **insurance**[2] for you and your family. In addition, I think you ought to **update**[3] your home insurance to make sure the **coverage**[4] is **adequate**[5].

In terms of[6] investments, why don't you speak to a qualified **financial**[7] advisor? I hear Tompkins Financial Consultancy is famous for its investment advice; they have an excellent **reputation**[8]. Personally, if I were you, I would put some money into the Indian and Brazilian markets right now. However, I really think the best thing to do is to talk to someone from Tompkins.

I hope my advice is helpful to you. Please don't hesitate to contact me again if you have any more questions.

Yours truly,
Donald

中译

收件人：特蕾莎·郭
发件人：唐纳德·杰克逊
主　题：保险与投资的相关建议

亲爱的特蕾莎：

关于你近日寄来的电子邮件向我询问保险与金融投资一事，最重要的是，我建议你为自己及家人购买个人寿险。此外，我想你应该更新房屋保险，以确保保险理赔范围完整无缺。

在投资方面，你何不找位合格的投资顾问谈谈？我听说汤普金斯财务顾问公司因投资咨询而闻名，该公司有绝佳的声誉。就个人而言，如果我是你，我现在就会把一些钱投资到印度和巴西市场。不过，我真的觉得最好的方法还是跟汤普金斯的顾问谈谈。

希望我的建议对你有帮助。你若有任何其他问题，请不要犹豫，尽管与我联系。

唐纳德 敬上

Vocabulary and Phrases

1. **investment** [ɪnˋvɛstmənt] *n.* 投资
   invest [ɪnˋvɛst] *vi.* & *vt.* 投资
   invest (sth) in...　（将某物）投资在……
   例: Now is not a good time to invest in the property market.
   （现在不是投资房地产市场的好时机。）
   Kevin invested his life savings in the stock market.
   （凯文把毕生积蓄投资在股票市场。）
   The company suffered huge losses due to a bad investment.
   （那家公司因投资不当而遭受巨大损失。）

2. **insurance** [ɪnˋʃurəns] *n.* 保险（不可数）
   insure [ɪnˋʃur] *vt.* & *vi.* 为……投保
   insure (sb / sth) against theft / fire...
   （为某人 / 某物）投保盗窃险 / 火险...
   例: Do you have insurance on your house and its contents?
   （你为房子和房内财物投保了吗？）
   It is wise to insure yourself against chronic diseases.
   （为自己投保慢性疾病险是明智的。）
   * chronic [ˋkrɑnɪk] *a.* 慢性的
   acute [əˋkjut] *a.* 急性的

3. **update** [ˌʌpˋdet] *vt.* 更新 & [ˋʌpˌdet] *n.* 更新
   update sb on sth
   向某人提供某事的最新消息
   a news update　新闻快报
   例: I called my boss to update him on the development of the project.
   （我打电话向老板报告这个项目的最新进展。）

4. **coverage** [ˋkʌvərɪdʒ] *n.* 保险理赔范围
   cover [ˋkʌvɚ] *vt.* 为……保险
   例: Are you fully covered for fire and theft?
   （你是否保了火灾和盗窃全险？）

5. **adequate** [ˋædəkwət] *a.* 妥当的
   be adequate for...　合乎……的需要
   例: The space available is not adequate for our needs.
   （现有的空间无法满足我们的需求。）

6. **in terms of...**　就……而言（此处 terms 常用复数）
   例: My last job was great in terms of salary, but it had its disadvantages.
   （就薪资而言，我上一份工作不错，但却有缺点。）

7. **financial** [faɪˋnænʃəl] *a.* 财务的

8. **reputation** [ˌrɛpjəˋteʃən] *n.* 声誉
   earn / establish / build a reputation
   赢得 / 建立声誉
   例: Mary soon earned a reputation as a first-class cook.
   （玛丽不久就赢得了一级厨师的荣誉。）
   This is a great opportunity to build our company's reputation.
   （这是个建立本公司声誉的好机会。）

Business Writing Exercises

请按括号中的提示将下列句子译成英文。

1. 我建议玛丽休几天假。（take a few days off）
   ____
2. 你何不做一个备份呢？（make a backup copy）
   ____
3. 北投以其温泉和餐厅而闻名。（hot springs）
   ____
4. 如果我是你的话，我会向你的同事道歉。（apologize to your colleague）
   ____
5. 就票房而言，这部电影极为成功。（the box office）
   ____

Unit 1

Making an Invitation　提出邀请

Basic Structure　基本结构

1. 说明活动内容并提出邀请。(Are you interested in...?)
2. 说明活动时间及地点。
3. 提醒对方回复是否要参加。(Please let me know if you'd like to attend...)

- 此类邮件旨在邀请对方参加活动，内容除了活动的相关信息外，也可以用"We would be honored if you could attend."（如果您能出席，我们将倍感荣幸。）"I really hope you can make it."（我真的希望你能来。）"I would like to invite you to attend..." "I was wondering if you would like to come to..."之类的话来表示期待对方的出席。

Sentence Patterns 写作句型 13

I would like to invite you to V / Would you like to V?
我想邀请你…… / 你想……吗?

- I would like to invite you to attend a party on Friday night.
  我想邀请你参加周五晚上的派对。
- Would you like to have dinner with me this weekend?
  这个周末你想跟我共进晚餐吗?

I was wondering if you would like to V / Are you interested in + N/V-ing?
我想知道你是否愿意…… / 你有兴趣……吗?

- I was wondering if you would like to visit our factory sometime.
  我想知道你是否想找个时间参观本工厂。
- Are you interested in going golfing in the near future?
  你最近有兴趣去打高尔夫球吗?

Let's V / Do you feel like + V-ing?
咱们去……吧。 / 你想不想……?

- Let's get together for lunch sometime this week.
  这星期咱们找时间共进午餐吧。
- Do you feel like going out for a walk after dinner?
  你晚餐后想出去走走吗?

You are formally invited to V / Please accept my / our invitation to V
我们正式邀请你…… / 请接受我(们)的邀请……

- You are formally invited to attend the company's annual meeting.
  我们正式邀请你参加本公司的年度会议。
- Please accept our invitation to the reception at the Lakeside Hotel.
  请接受我们的邀请出席在湖滨饭店举行的招待会。

I / We would be honored if you could V / We request the pleasure of your presence at...
如果您能……我(们)会倍感荣幸 / 请赏光出席……

- We would be honored if you could attend the wedding of our daughter.
  如果您能参加我们女儿的婚礼,我们会倍感荣幸。
- We request the pleasure of your presence at our anniversary party.
  请赏光出席我们的周年派对。

Model E-mail 电子邮件范例  13

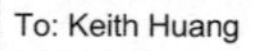

To: Keith Huang

From: Anne Crenshaw

Subject: New Product **launch**[1]

Dear Keith:

I would like to invite you to **attend**[2] the launch of our new product on June 16 at 2 p.m. at the Bellevue Hotel. In addition, I was wondering if you would like to come to a **reception**[3] after the official launch. The reception, including **finger food**[4] and **beverages**[5], will be held from 3:30 p.m. to 6 p.m.

I really hope you can **make it**[6], Keith. Also, do you feel like having lunch sometime next week? If you're available, let's go to that new Turkish restaurant downtown. Please let me know.

Best regards,
Anne

To: Jasmine Rush

From: Tara Jenkins

Subject: Our 10th **anniversary**[7]

Dear Ms. Rush:

Our company is celebrating its 10th anniversary with a **formal**[8] dinner on November 12. You are formally invited to the dinner, which will begin at 7 p.m. at the Retinue Restaurant, 1036 E. Broadway Avenue.

We would be **honored**[9] if you could attend. Please **RSVP**[10] by November 8. Thank you, and we look forward to seeing you there.

Sincerely,
Tara Jenkins for
Mr. Brandon Xu
President, Makeover International

中译

收件人：基思·黄
发件人：安妮·克伦肖
主　题：新品上市

亲爱的基思：

我想邀请你参加本公司6月16日下午两点在贝勒维饭店举行的新品上市发布会。此外，我想知道你是否愿意在正式发布会后参加招待会。招待会将于下午3点半到6点间举行，其间提供小点心和饮料。

基思，我真的希望你能来。此外，下星期你想不想找时间一起吃顿午饭？如果你有空的话，咱们可以去那家在市中心新开的土耳其餐厅。请回信告知。

诚挚的祝福

安妮

中译

收件人：洁思敏·拉什
发件人：塔拉·詹金斯
主　题：我们的10周年纪念

亲爱的拉什小姐：

本公司将于11月12日举办正式晚宴，庆祝公司成立10周年。本公司正式邀请您参加这个晚宴，晚宴将于晚上7点开始，地点在百老汇东大街1036号的雷特尤餐厅。

如果您能出席，我们将倍感荣幸。敬请在11月8号前回信告知。谢谢您，我们期待与您见面。

塔拉·詹金斯代表国际化妆公司总裁
布兰登·许　先生　敬上

Vocabulary and Phrases

1. **launch** [lɔntʃ] *n.* & *vt.* (产品的)上市；发起(活动)

   例: The official launch date is May 1.
   (正式发行日是 5 月 1 日。)

   The company is scheduled to launch the new product by next October.
   (该公司预计明年 10 月以前要推出那款新产品。)

   The marketing department will launch an ad campaign to promote our new product.
   (营销部将进行广告活动以促销新产品。)

2. **attend** [əˋtɛnd] *vt.* 参加
   attend a meeting / wedding / funeral
   参加会议/婚礼/葬礼

   例:Mike attended a 4-day conference on telecommunications.
   (迈克参加了为期 4 天的电信研讨会。)

3. **reception** [rɪˋsɛpʃən] *n.* 招待会，欢迎会
   a wedding reception 结婚喜宴

4. **finger food** (手抓)小点心

5. **beverage** [ˋbɛvərɪdʒ] *n.*
   (除水以外的)饮料

6. **make it (to...)** 出席；赶到(……)

   例: I won't be able to make it to your party on Saturday.
   (我无法出席你星期六的派对。)

   The train leaves in five minutes. We'll never make it.
   (火车 5 分钟后就要开了。我们赶不上了。)

7. **anniversary** [ˌænəˋvɜsərɪ] *n.*
   周年纪念日

8. **formal** [ˋfɔrml̩] *a.* 正式的
   formally [ˋfɔrml̩ɪ] *adv.* 正式地

   例: I kept the tone of the letter formal and businesslike.
   (我使这封信保持正式商务的语气。)

9. **honored** [ˋɑnəd] *a.* 荣幸的
   be / feel honored to V ……感到荣幸

   例: I was honored to have been mentioned in the professor's speech.
   (教授在演讲中提到了我，让我倍感荣幸。)

10. **RSVP** 敬请回复(请柬用语，源自法语，是 Répondez S'il Vous Plaît 的缩写，直译成英语为 Please reply.)

Business Writing Exercises

请按括号中的提示将下列句子译成英文。

1. 我想邀请你做我们演讲比赛的评委。(a judge of our speech contest)
   ____

2. 我很想知道你是否对成为歌手感兴趣。
   ____

3. 你周末想去游泳吗?
   ____

4. 恐怕我无法赶去参加你的生日派对。(make it to...)
   ____

5. 请赏光出席我们的结婚周年纪念。(at our wedding anniversary)
   ____

Unit 2

Responding to Invitations 回复邀请

Basic Structure 基本结构

1. 感谢对方的邀请，并表示当日是否可以出席。（It's kind of you to think of me.）
2. 若不确定是否能出席，告知对方将尽快回复。（I'm not sure of my plans for...）
3. 若确定无法出席，则告知对方原因。（I have another commitment for that day.）（I'm sorry, but I'm otherwise engaged.）

- 收到邀请时，无论是否参加都必须回复。若不能前往，最好说明理由。
- 如要对对方的邀请表示感谢，可以使用“Thank you very much for your invitation.”（非常感谢你的邀请。）之类的句子。
- 如想婉拒邀请，可以使用“I'd love to, but I can't...”（我很乐意，但我不行……）之类的句子来表示自己无法到场。

Sentence Patterns 写作句型 14

Thank you very much. I'd love to V / That's kind of you to V
非常感谢，我乐意…… / 你……真是太好了。

- Thank you very much for your invitation. I'd love to come.
  非常感谢你的邀请，我很乐意去。
- That's kind of you to invite me. Thank you.
  谢谢，你能邀请我真是太好了。

(Sth) sounds great. / (Sth) is a great idea.
（某事）听起来不错。 / （某事）是个好主意。

- Dinner on the 22nd sounds great. Sure, thanks.
  22 号的宴会听起来不错。没问题，谢谢。
- Visiting a museum on Saturday is a great idea.
  星期六去参观博物馆真是个好主意。

I'm not sure of my plans for (day or date) right now. / I'll let you know soon about (sth).
我目前不太确定（某天或某日期）的安排。 / 我会尽快告知你（某事）。

- I'm sorry, but I'm not sure of my plans for Saturday right now.
  抱歉，我目前不太确定星期六的安排。
- I'll let you know soon about the party.
  我会尽快告知你有关派对的事。

Thank you, but unfortunately I'm unable to make it to ... / I'm really sorry, but I have other plans for (day or date).
谢谢，但不巧的是我无法参加…… / 非常抱歉，我（某天或某日期）已安排了其他事情。

- Thank you for your invitation, but unfortunately I'm unable to make it to lunch on Friday.
  感谢你的邀请，但不巧的是星期五我不能与你共进午餐。
- I'm really sorry, but I have other plans for Oct. 22. Would it be possible to meet for lunch on the 6th of October?
  非常抱歉，我 10 月 22 日已安排了其他事情。有没有可能 10 月 6 日一起吃个午餐？

Regretfully, I'm not free on (day or date). / I'd love to, but I can't on (day or date).
遗憾的是，我（某天或某日期）没空。 / 我很乐意，但我（某天或某日期）不行。

- Regretfully, I'm not free this Sunday.
  遗憾的是，我这星期日没空。
- I'd love to attend your party, but I can't on the 14th.
  我很乐意参加你的派对，但我 14 日那天不行。

Model E-mail 电子邮件范例 14

To: Marnie Cartwright

From: Arlene Platt

Subject: Dinner at the Emporium

Dear Marnie:

It was very nice to get your recent invitation. That's very kind of you to **think of**[1] me. Thank you. Dinner on April 10 sounds great, and having it at The Emporium Restaurant is a great idea, I think. I'll see you on the 10th. Take care.

Best regards,
Arlene

中译

收件人：马妮·卡特莱特
发件人：艾琳·普拉特
主 题：恩波里安餐厅的晚餐聚会

亲爱的马妮：

能收到你日前的邀请真好。你能想到我真是太好了，谢谢。4 月 10 日的晚餐聚会听起来不错，而且我觉得在恩波里安餐厅举办宴会这个点子真棒。期待 10 日那天与你相见。保重。

诚挚的祝福

艾琳

To: Tony Ma

From: Deborah Wang

Subject: **Reception**[2]

Hi Tony:

Thanks for inviting me to the reception to celebrate the opening of your new office. **I'm** not **sure of**[3] my plans for next Thursday afternoon. I'll let you know soon about the reception.

Deborah

中译

收件人：托尼·马
发件人：德博拉·王
主 题：招待会

你好，托尼：

感谢你邀请我去招待会，以庆祝你的新公司开张。我不太确定下星期四下午的安排，我会尽快告知你招待会的事。

德博拉

To: Kelly West

From: Patty Zhu

Subject: **Retirement**[4] dinner

Dear Kelly:

It was great to get your e-mail regarding the retirement dinner for your husband. Thank you, but unfortunately I'm unable to make it to the dinner. **Regretfully**[5], I'm not **free**[6] on the 4th, as I have another **commitment**[7] for that date. **Congratulations**[8] to your husband, Ray, on his retirement!

Sincerely,
Patty

中译

收件人：凯莉·威斯特
发件人：帕蒂·朱
主 题：退休晚宴

亲爱的凯莉：

能收到关于你丈夫退休晚宴的电子邮件感觉真好。谢谢，但不巧的是，我无法参加当日晚宴。很遗憾，我 4 日没空，因为那天我已经先答应别人的约会了。祝贺你丈夫雷退休了！

帕蒂 敬上

Vocabulary and Phrases

1. **think of / about + N/V-ing** 想到……
   例: When I said that, I wasn't thinking of anyone in particular.
   （我说那话时，并没有特别想到谁。）
2. **reception** [rɪˈsɛpʃən] *n.* 欢迎会，招待会
3. **be sure of sth** 确信某事
   例: Kate wasn't completely sure of coming to the party tonight.
   （凯特不完全确定今晚会不会来派对。）
4. **retirement** [rɪˈtaɪrmənt] *n.* 退休
   retire [rɪˈtaɪr] *vi.* 退休
   retire from... 从……退休
   例: Mary was forced to retire from teaching due to ill health.
   （玛丽因为健康不佳而被迫从教职退休。）
5. **regretfully** [rɪˈgrɛtfəlɪ] *adv.* 遗憾地
6. **free** [fri] *a.* 有空的（= available）
   例: I won't be free all day today because of such a heavy workload.
   （工作量那么大，因此我今天整天都没空。）
7. **commitment** [kəˈmɪtmənt] *n.* 已承诺（或同意）的事
   例: Career women have to juggle work with their family commitments.
   （职业妇女经常得两头兼顾，既要工作又要照管家庭。）
   * juggle [ˈdʒʌgl̩] *vt.* 尽力同时应付（两个以上重要的工作或活动）
8. **congratulations** [kənˌgrætʃəˈleʃənz] *n.* 祝贺；道喜（常用复数，与介词 on 连用）
   congratulate [kənˈgrætʃəˌlet] *vt.* 祝贺；道喜
   congratulate sb on sth
   就某事恭喜某人
   例: Congratulations on your engagement!
   （恭喜你订婚了！）
   I congratulated Tom on his test results.
   （我就汤姆考试的结果向他道贺。）

Business Writing Exercises

请按括号中的提示将下列句子译成英文。

1. 你表示愿意帮忙真是太好了，不过我自己可以应付得来。（manage it myself）
   ____
2. 我不能留下来吃晚餐，但喝杯茶听起来不错。
   ____
3. 你有必要每天运动。（It's necessary for you to...）
   ____
4. 非常抱歉，我已另有安排。（have other plans）
   ____
5. 恭喜你最近荣升为经理！（promotion to manager）
   ____

Unit 1

Inquiring about Product Information
询问产品情况

Basic Structure 基本结构

1. 表明来意后，简单说明公司背景及产品属性。
2. 要求获得该公司产品目录（catalog）或广告小册子（brochure）。
3. 表明未来愿意合作的意愿。（I look forward to possibly doing business with you in the near future.）

- 这类电子邮件的目的是向对方索取产品资料，因此可以用“We are considering switching suppliers, so we are now in the market for various components.”（本公司正在考虑更换供应商，所以现正有意购买各种不同的零件。）之类的说法来让对方了解索取的动机。
- 写作时使用主动语态可以让内容更生动，更具个人化，且更加有趣。避免使用被动语态。

Sentence Patterns 写作句型 15

I'm writing to inquire about... / I'd like to request some information about / on...
我来信询问…… / 我想请求获得……的相关资料。

- I'm writing to inquire about your products.
  我来信询问贵公司的产品。
- I'd like to request some information on your cell phones.
  我想请求获得贵公司手机产品的相关资料。

Could you send me a catalog / catalogue of...? / Do you have any brochures of...?
你可否寄一份……的目录给我？ / 你有……的小册子吗？

- Could you send me a catalog of your computer accessories?
  你可否寄一份电脑配件的目录给我？
- Do you have any brochures of your power tools?
  你们有电动工具的小册子吗？
  * brochure [broˋʃur] *n.* 小册子

I would be grateful if you could... / I would appreciate it if you could...
如果您能……我将不胜感激。

- I would be grateful if you could e-mail me details about the wireless keyboards.
  如果您能用邮件发给我无线键盘的详细资料，我将不胜感激。
- I would appreciate it if you could send me the information I requested.
  如果您能发给我所要求的资料，我将不胜感激。

We are in the market for... / We are considering buying...
本公司有意购买…… / 本公司正在考虑购买……

* be in the market for sth 有意购买某物

- We are in the market for a new factory to manufacture our products.
  本公司有意购买新厂房，以大量制造产品。
- We are considering buying some new office equipment.
  我们正在考虑购买一些新的办公设备。

We look forward to receiving... / We eagerly await your reply regarding...
我们期待收到…… / 我们热切期待收到您关于……的答复。

- I look forward to receiving the information about your goods.
  我期待收到有关贵公司货物的资料。
- We eagerly await your reply regarding the automotive parts.
  我们热切期待收到您关于汽车零件的答复。

Model E-mail 电子邮件范例 15

To: Top-Notch Electronics Co., Ltd.

From: Ian Xiao

Subject: Components for consumer electronics

Dear Top-Notch Electronics:

I'm writing to **inquire about**[1] your electronic **components**[2]. We are a company based in California that **manufactures**[3] consumer electronics. I read an article about your company in the business **section**[4] of our local newspaper. Could you send me a catalog of your components **related to**[5] MP3 players and digital cameras? I would appreciate it if you could send me a catalog or brochure by e-mail **at your earliest convenience**[6]. We are considering **switching**[7] **suppliers**[8], so we are now **in the market**[9] for **various**[10] components.
I look forward to receiving the information about your parts and possibly doing business together **in the near future**[11].

Yours truly,
Ian Xiao
Supply Manager
Power Electronics Inc.

中译

收件人：顶尖电子有限公司
发件人：伊恩・肖
主　题：消费性电子产品零件

顶尖电子公司亲爱的同仁们好：

我来信旨在询问贵公司电子零件一事。本公司总部位于加州，主要生产消费性电子产品。我在本地报纸的商业专栏看到一篇关于贵公司的文章。贵公司可否寄给我一份 MP3 播放器和数码相机的零件的目录？方便时若能尽早将目录或小册子用电子邮件发送给我，本人将不胜感激。本公司正在考虑更换供应商，所以现正有意购买各种不同的零件。

我期待收到有关贵公司零件的资料，也期待在不久的将来可能与贵公司做生意。

顶尖电子有限公司　供应经理
伊恩・肖　敬上

Vocabulary and Phrases

1. **inquire about...** 询问有关……
   inquire [ɪnˈkwaɪr] *vi.* 询问(= enquire)
   例: I am writing to inquire about your advertisement in *The China Post*.
   (我来信是想询问贵公司在《中国邮报》上刊登的广告事宜。)
2. **component** [kəmˈponənt] *n.* 组成部分;零件
   car components 汽车零件
   = car parts
   例: Trust is a vital component in any relationship.
   (在任何关系中,信任都是一个至关重要的因素。)
   * vital [ˈvaɪtḷ] *a.* 极重要的,不可或缺的
3. **manufacture** [ˌmænjəˈfæktʃɚ] *vt.* 大批生产
   manufacturer [ˌmænjəˈfæktʃərɚ] *n.* 制造商,厂商
   例: Jack works for a company that manufactures car parts.
   (杰克任职于一家生产汽车零件的公司。)
4. **section** [ˈsɛkʃən] *n.* 部分
   the sports section of the newspaper 这份报纸的体育版
   例: We'll discuss the issue in more detail in section four of the proposal.
   (我们会在这份提案的第四部分更详细地讨论这个问题。)
5. **related** [rɪˈletɪd] *a.* 有关的
   be related to... 与……相关
   例: The murder <u>was</u> reportedly <u>related to</u> drugs.
   = The murder <u>was</u> reportedly <u>connected with</u> drugs.
   (据报道,这起谋杀案跟毒品有关。)
6. **at one's earliest convenience** 某人方便时尽早
   例: Call me at your earliest convenience.
   (您方便时尽早给我来电。)
7. **switch** [swɪtʃ] *vt. & vi.* 改变;转换
   例: The dates of the two exams have been switched.
   (这两门考试改期了。)
   We're in the process of switching over to a new computer system.
   (我们正在转换使用新的电脑系统。)
8. **supplier** [səˈplaɪɚ] *n.* 供应商
9. **in the market for sth** 有意购买某物
   例: I'm not in the market for a new car at the moment.
   (我此刻还不想买新车。)
10. **various** [ˈvɛrɪəs] *a.* 各种各样的
   = a wide variety of...
   = a wide range of...
   例: As I see it, there are various options open to you.
   (依我看,你有很多选择。)
   * option [ˈɑpʃən] *n.* 选择
11. **in the near future** 在不久的将来

Business Writing Exercises

请按括号中的提示将下列句子译成英文。

1. 我来信是询问你们的新产品及其价格。

   ____

2. 你能否给我寄一份你们的电器目录?(electrical appliances)

   ____

3. 如果你能随时给我提供消息,我将不胜感激。(keep me informed / posted)

   ____

4. 我们是一家总部设在马来西亚的家具制造商。

   ____

5. 我们有意购买 30 棵圣诞树。

   ____

Unit 2

Requesting a Quotation / Quote
请求对方报价

Basic Structure 基本结构

1. 感谢对方公司寄来的目录或小册子。说明本公司目前正在询求某材料的报价。
2. 向对方说明所要求报价的物品数量、尺寸、编号。
3. 询问订购相关事宜（如：最低订购量、折扣等）

- 一封完整的询价信应有以下要点：
  1. 扼要说明需求。
  2. 解释货品的用途，如此供应商才能提供符合需要的报价。
  3. 说明所需数量。
  4. 说明何时交货。
  5. 询问报价包含哪些费用，另外还要问清楚折扣、包装、运费及付款条件等内容。
- 请求报价时，别忘了要问清楚价格是否包含运费和保险费等。否则可能会因为供应商的报价不明确而造成双方的误会。

Sentence Patterns 写作句型 16

Could you provide me with a quotation for...? / I would like to have a quote on...
你能否就……给我个报价？ / 我想请你给……报个价。

- Could you supply me with a quotation for 100,000 gaskets?
  你能否就 10 万个垫圈报个价？
  * gasket [ˈgæskɪt] *n.* 垫圈
- I would like to have a quote on 50,000 desks.
  我想请你给 5 万张书桌报个价。

We are seeking a price quote for... / Please quote us a (firm) price for...
我们在征询……的报价。 / 请给我们报出……（确定的）价格。

- We are seeking a price quote for rubber mats.
  我们在征询橡胶垫的报价。
- Please quote us a firm price for 250,000 printer cables.
  请给我们报出 25 万条打印机电缆的确切价格。

We are now soliciting quotes for... / Our company is currently seeking prices for...
我们正在征询……的报价。 / 本公司目前正在询问……的价格。

- We are now soliciting quotes for coil springs.
  我们正在征询线圈弹簧的报价。
  * solicit [səˈlɪsɪt] *vt.* 征求
- Our company is currently seeking prices for plastic covers.
  本公司目前正在询问塑料盖的价格。

What's your best price for...? / Do you offer discounts on bulk orders for...?
贵公司……的最优惠价格是多少？ / 贵公司针对大量订购……有折扣吗？

- What's your best price for 500 glass panels?
  贵公司 500 块玻璃嵌板的最优惠价格是多少？
  * panel [ˈpænl̩] *n.* （玻璃）嵌板
- Do you offer discounts on bulk orders for screws?
  贵公司针对螺丝钉的大宗订单有折扣吗？
  * bulk [bʌlk] *n.* 大量
  a bulk order 大宗订单

What is the minimum order for...? / Is there a minimum requirement for orders on...?
……的最低订购量是多少？ / 请问……的最低订购量要求是多少？

- What is the minimum order for aluminum cans?
  铝罐的最低订购量是多少？
- Is there a minimum requirement for orders on wooden pegs?
  请问木衣夹的最低订购量要求是多少？
  * peg [pɛg] *n.* 晾衣夹子

Model E-mail 电子邮件范例  16

To: Larry Crown, Landover **Tubing**[1] and **Pipe**[2] Co.

From: Tanya Morgan, Midtown **Landscaping**[3] Ltd.

Subject: Plastic tubing

Dear Mr. Crown:

Thank you for sending me your brochure on the different types of tubing and pipes your company produces. We are seeking a price **quote**[4] on tubing. Could you provide me with a quotation on 1,500 rubber tubes, each cut into 50 cm sections? In your brochure, the **reference**[5] number on the tubing is QT-3287. Is there a **minimum**[6] requirement for orders of this material? Also, do you offer discounts on bulk orders? Please let me know soon so I can discuss this with our company's general manager.

Yours truly,
Tanya Morgan
Midtown Landscaping Ltd.

中译

收件人：兰德佛管材管件公司，拉里·克朗
发件人：市中心景观美化股份有限公司，
坦尼娅·摩根
主　题：塑料管

亲爱的克朗先生：

感谢你发给我贵公司生产的不同管料的广告小册子。本公司正在征询管料的报价。你能否给我提供 1500 条橡皮管的报价？每条橡皮管要切成 50 厘米长。在贵公司的广告小册子中列出了该管料的参考编号为 QT-3287。这类材料的最低订货量要求是多少？此外，大量订购的话贵公司有折扣吗？请尽快告知我，这样我才能和我们总经理讨论相关事宜。

市中心景观美化股份有限公司
坦尼娅·摩根　敬上

To: Randy Schnell, Dart Industries

From: Sheldon Dong

Subject: Prices on **concrete**[7] and steel

Dear Mr. Schnell:

I am the **procurement**[8] manager for a company called Talon Construction. We are now **soliciting**[9] quotes for concrete and steel. In the **attachment**[10] you will find the **exact**[11] **specifications**[12] regarding concrete and steel that we require. What's your best price on this type of concrete and steel?

Best regards,
Sheldon Dong,
Talon Construction

写作工具箱

公司属性常见缩写介绍：（注意以下缩写词的发音仍须念出全名，如：Co. 要念成 company）。

Co. (= company) 公司

Ltd. (= limited) 股份有限公司（用于英国公司或商行名称之后）

Inc. (= incorporated) 公司（美国用法，置于公司名称之后）

Corp. (= corporation)（大）公司

中译

收件人：达特工业，兰迪·施内尔
发件人：谢尔顿·董
主　题：钢筋混凝土价格

亲爱的施内尔先生：

本人是一家名为塔隆建筑公司的采购经理。本公司正在征询钢筋混凝土的报价。你可在附件中见到本公司所需钢筋混凝土的具体规格。贵公司这类钢筋混凝土的最优惠价格是多少？

诚挚的祝福

塔隆建筑公司
谢尔顿·董

Vocabulary and Phrases

1. **tubing** [ˈtjubɪŋ] *n.* 管料；金属管（不可数）
   a piece of copper tubing　一根铜管
   tube [tjub] *n.*（金属、塑料、橡胶）管子（可数）
2. **pipe** [paɪp] *n.* 管子
   a water pipe　水管
3. **landscaping** [ˈlændˌskepɪŋ] *n.* 景观美化
   landscape [ˈlændˌskep] *vt.* 对……做景观美化 & *n.* 乡村风景（画）
4. **quote** [kwot] *n.* (=quotation [kwoˈteʃən]) & *vt.* 报价
   例: Ask the architect to give you a written quotation for the job.
   （要求建筑师给你一份这项工作的书面报价。）
   They quoted us $200 for installing a shower unit.
   （他们向我们报价 200 美元安装淋浴设备。）
5. **reference** [ˈrɛfərəns] *n.*（为方便查询所用的）编号
6. **minimum** [ˈmɪnɪməm] *n.* 最少量 & *a.* 最少的
   maximum [ˈmæksɪməm] *n.* 最大量 & *a.* 最多的
   例: The class needs a minimum of 20 students to continue.
   （这个班最少要有 20 个学生才能继续下去。）
   The minimum age for entering most nightclubs is 18.
   （进入大部分夜店的最低年龄限制是 18岁。）
7. **concrete** [ˈkaŋkrit] *n.* 混凝土
8. **procurement** [prəˈkjʊrmənt] *n.* 采购（不可数）
   procure [prəˈkjʊr] *vt.*（设法）获得
   例: The man was accused of procuring weapons for terrorists.
   （那名男子被控为恐怖分子购买武器。）
9. **solicit** [səˈlɪsɪt] *vt.* 请求给予（援助、钱或信息）
   例: They sent sales representatives abroad to solicit business.
   （他们派业务员出国招揽生意。）
10. **attachment** [əˈtætʃmənt] *n.*（电子邮件发送的）附件
11. **exact** [ɪgˈzækt] *a.* 精准的
    例: The woman gave an exact description of the attacker.
    （那名女子精确描述出攻击者的样貌。）
    Gary is in his early thirties—thirty-two, to be exact.
    （加里 30 出头——准确地说是 32 岁。）
12. **specification** [ˌspɛsəfəˈkeʃən] *n.* 规格（简称 spec [spɛk]）
    例: We want the machine manufactured exactly to our specifications.
    （我们希望这台机器能完全按照我们要求的规格生产。）

Business Writing Exercises

请按括号中的提示将下列句子译成英文。

1. 你能否就 1000 把雨伞给我提供一个报价？
   ____
2. 请就下列项目向我们报价。
   ____
3. 你们 500 条牛仔裤的最优惠价格是多少？
   ____
4. 你们针对大宗订单有什么折扣吗？（bulk orders）
   ____
5. 要求送货的最低订购量是多少？（...for delivery）
   ____

Unit 3

Responding to Inquiries 回应询问

Basic Structure 基本结构

1. 感谢来信询问商品，并附上要求的商品目录/价目表。(Please find our catalog and price list attached to this e-mail.)
2. 提醒优惠的时限与最低订购数量（the minimum order）。
3. 欢迎再次询问相关细节，并提供联系方式。

- 这类电子邮件的目的是针对买方进行报价，因此内容可包括：
  1. 对询问表示感谢。
  2. 提供价格、折扣和付款条件的细节。
  3. 说明价格是否包含包装、运费或保险。
  4. 约定交货日期。
  5. 报价的有效期限。
  6. 希望对方能接受此报价。
- 如要留下其他的联系方式，可以使用“If you have any further questions, please don't hesitate to contact me at orders@ezfit.com.”（如果您有任何更进一步的问题，请不要犹豫与本人联系，电子邮箱是 orders@ezfit.com。）之类的句子向对方说明。

Please find our (brochure / catalog / price list) attached to this e-mail. / We are including a copy of our (brochure / catalog / price list) for your consideration.

请查收此封电子邮件附件内的（广告小册子 / 目录 / 价目表）。 / 我们附上（广告小册子 / 目录 / 价目表）以供参考。

- Please find our brochure on ink cartridges attached to this e-mail.
  请查收此封电子邮件附上的墨盒广告小册子。
- We are including a copy of our bicycle parts catalog.
  兹附上本公司自行车零件的目录副本。

We are pleased to provide you with our (catalog / price list / quotation). / As you requested, here is our (catalog / price list / quotation).

我们很乐意给您提供我们的（目录 / 价目表 / 报价单）。 / 按照您的要求，这是我们的（目录 / 价目表 / 报价单）。

- We are pleased to provide you with our price list on testing machines.
  我们很乐意给您提供我们试验机的价目表。
- As you requested, here is our quotation for machine parts.
  根据您的要求，这是我们机器零件的报价单。

The quote is valid for (period of time). / This offer is good until (date).

本报价有效期限为（某段时间）。 / 此优惠价直到（某日期）。

- The quote is valid for three months.
  本报价有效期限为 3 个月。
- This special offer is good until August 31.
  本优惠价直到 8 月 31 号。

The quotation is based on a minimum order of (number) / We require a minimum order of (number).

本报价是基于（某数量）的最低订购量。 / 我们要求最低订购量为（某数量）。

- The quotation is based on a minimum order of 1,000 pieces.
  本报价是基于 1000 件的最低订购量。
- We require a minimum order of 2,000 items.
  我们要求最低订购量为 2000 件。

If you have any further questions, please don't hesitate to contact us. / I'd be happy to speak to you further if you have any questions.

如果您有任何更进一步的问题，请不要犹豫与我们联系。 / 如果您有任何问题，我很乐意与您进一步详谈。

- If you have any further questions, please don't hesitate to contact me at orders@ezfit.com.
  如果您有任何更进一步的问题，请不要犹豫与本人联系，电子邮箱是 orders@ezfit.com。
- I'd be happy to speak to you further if you have any questions. My phone number is 0918-589-123.
  如果您有任何问题，我很乐意与您进一步详谈。我的电话号码是 0918-589-123。

Model E-mail 电子邮件范例 17

To: Jackie Wellman

From: John Zhou

Subject: **Inquiry**[1] about **ball bearings**[2]

Dear Ms. Wellman:

Thank you for your recent e-mail inquiring about the ball bearings **manufactured**[3] by our company. Please find our brochure attached to this e-mail. Also, as you requested, here is our price list based on your specifications. Please **note**[4] that this offer is **valid**[5] for 90 days only. I should also remind you that the quotation **is based on**[6] a minimum order of 1,000 pieces. If you have any further questions, please don't hesitate to contact me at John@highrollers.com. If you prefer to **place an order**[7] over the telephone, please contact us at 866-2-8322-9874. One of our sales staff would be happy to speak to you further on this matter.

Yours truly,
John Zhou
Sales manager
High Rollers International

中译

收件人：杰姬·韦尔曼
发件人：约翰·周
主　题：询问有关滚珠轴承一事

亲爱的韦尔曼女士：

感谢您近期发来电子邮件询问本公司生产的滚珠轴承。请查收此封电子邮件附件内的广告小册子。此外，根据您的要求，这里有一份按照您要求的规格所列的价格清单。请注意，本优惠有效期限只有 90 天。我还要提醒您，此份报价单是按最低订购量 1000 件为基准。您若有任何更进一步的问题，请不要犹豫与我联系，邮箱是 John@highrollers.com。如果您较喜欢用电话下单，请拨打 866-2-8322-9874 与我们联系。本公司业务员将很乐意与您详谈此事。

高辊国际有限公司　销售经理
约翰·周　敬上

Vocabulary and Phrases

1. **inquire** [ɪnˈkwaɪr] *vi.* 询问(= enquire)
   inquire [ɪnˈkwaɪr] *vi.* 询问(与介词 about 连用)
   make a few / some inquiries 问些问题
   inquire about... 询问……
   例: I'll have to make some inquiries and get back to you.
   (我得打听打听再回复你。)
   I'm calling to inquire about the advertisement in the newspaper.
   (我来电是询问报纸上的那则广告。)
2. **ball bearing** [ˈbɔlˌbɛrɪŋ] *n.* 滚珠轴承
3. **manufacture** [ˌmænjəˈfæktʃɚ] *vt.* (使用机器大规模)制造,生产
   manufacturer [ˌmænjəˈfæktʃərɚ] *n.* 制造商,厂商
4. **note** [not] *vt.* & *n.* 注意(= notice [ˈnotɪs])
   please note / notice that... 请注意……
   note sth 注意某事物
   = notice sth
   = take notice of sth(notice 为名词)
   = take note of sth(note 为名词)
   例: Please note that the bill must be paid within thirty days.
   (请注意此账单须于 30 天内缴纳。)
   People are beginning to take note of Jane's talents as a writer.
   (大家开始注意到简的写作天分。)
5. **valid** [ˈvælɪd] *a.* 有效的
   be valid for + 一段时间 若干时间内有效
   例: The ticket is valid for three months.
   (这张票 3 个月内都有效。)
6. **be based on / upon...** 以……为基础
   例: Marital life should be based on mutual trust and understanding.
   (婚姻生活应该以互信和体谅为基础。)
   * marital [ˈmærətḷ] *a.* 婚姻的
   mutual [ˈmjutʃuəl] *a.* 相互的
7. **place an order for sth** 下订单购买某物
   例: I'd like to place an order for ten copies of this book.
   (我想下订单购买 10 册这本书。)

Business Writing Exercises

请按括号中的提示将下列句子译成英文。

1. 请查收附于此封电子邮件的本公司价目表。

   ____

2. 我们很乐意给您提供最新的目录。

   ____

3. 由于价格时有变动,本报价有效期限为 3 天。(Due to price fluctuations, the quote...)

   ____

4. 本报价以 600 台的最低订购量为基准。

   ____

5. 如果有任何问题产生,请不要犹豫,直接跟我们联系。(If any problem comes up,...)

   ____

Unit 1

Placing an Order 下订单

Basic Structure 基本结构

1. 说明欲订购的商品数量及型号。
2. 请卖方确认总价并提供买方交货地址。
3. 提供订单明细。

- 这类电子邮件的目的是下订单，因此开头无需使用客套话。
- 即使发件人的邮件地址有时会出现在签名档中，最好还是将其纳入电子邮件的内文里，并和产品型号一并以黑体表示，以方便对方阅读。
- 如需要在特定日期前送到，可以用“Please ensure delivery by November 4.”（请务必最晚于11月4日交货。）之类的句子来提醒对方最后的期限。

Sentence Patterns 写作句型 18

I'd like to place an order for (quantity) of (item). / Please accept this order for (quantity + item).
我想订（某数量）的（某商品）。/ 请接受此（某数量的商品）订单。

- I'd like to place an order for 1,500 gas cylinders.
  我想订购 1500 个煤气罐。
  * cylinder [ˈsɪlɪndɚ] *n.* 圆筒；气缸
- Please accept this order for 10,000 ring binders.
  请接受此 10000 份活页夹的订单。
  * ring binder [ˈrɪŋˌbaɪndɚ] *n.* 活页夹

Please find the details of our order (that is) attached. / I have attached a purchase order for...
兹附上我们订单的明细，请查收。/ 我已附上一份采购订单。

- Please find the details of our order for 15,000 units (that are) attached.
  兹附上我们 15000 组的订单明细，请查收。
- I have attached a purchase order for 250,000 silicon chips.
  我已附上一份 25 万个硅芯片的采购订单。

Could you please confirm that the price of (price amount) is correct? / As agreed to previously, I believe the amount is (price amount).
可否请您确认（某总价）价格无误？/ 按先前议定，我认为金额是（某总价）。

- Could you please confirm the price of $65,000 is correct?
  可否请您确认 65000 美元的价格无误？
- As agreed to previously, I believe the amount is $10,350.
  按先前议定，我认为金额是 10350 美元。

Please ship the items to the following address: (address). / We will accept delivery at this address: (address).
请将商品运送到下列地址：（某地址）。/ 本公司将在此地址接受交货：（某地址）。

- Please ship the items to the following address: 3598 Kingston Boulevard, Miami, Florida, 904.
  请将商品运送到下列地址：佛罗里达州迈阿密市，金士顿大道 3598 号，邮政编码 904。
- We will accept delivery at this address: #54 – 13098 Arrow Avenue, London, England, N10AX.
  本公司将在此地址接受交货：英国伦敦市，亚罗大街 54 - 13098 号，邮政编码 N10AX。

Please ensure delivery by (date). / We require the (items) to reach us by (date).
请务必最晚于（某日期）交货。/ 本公司要求（某商品）最晚在（某日期）送到。

- Please ensure delivery by November 4.
  请务必最晚于 11 月 4 日交货。
- We require the cables to reach us by January 27.
  本公司要求这些电缆最晚于 1 月 27 日送到。

Model E-mail 电子邮件范例 18

To: Ralph Henman, Wing Computer Products

From: Jack Fang, Beautiful Island Electronics

Subject: Order

Dear Mr. Henman:

I'd like to **place an order for**[1] 4,500 electric **switches**[2], model number JX-980-12. **In addition**[3], please accept our order for 1,500 capacitors, model number CP-765-44. Please find the details of the order attached to this e-mail. Could you please **confirm**[4] that the total price of $44,290 is correct? Please **ship**[5] the items by November 4 to the following address:

Beautiful Island Electronics
258 Zhongxiao East Rd., Sec. 3
Taipei, Taiwan, 106

Sincerely,
Jack Fang,
Purchasing manager,
Beautiful Island Electronics

[Attachment]
Wing Computer Products Purchase order

Quantity	Description	**Unit Price**[6]	Total
4,500	Electric switches (model number JX-980-12)	$8.65	$38,925.00
1,500	Capacitors (model number CP-765-44	$0.75	$1,125.00
Subtotal[7]			$40,050.00
Sales Tax (10%)			$4,050.00
Shipping & **Handling**[8]			$190.00
Total			$44,290.00

中译

收件人：展翼电脑产品公司，拉尔夫・赫曼
发件人：美丽岛电子公司，杰克・方
主　题：订单

亲爱的赫曼先生：

我想订购4500组型号为JX-980-12的电源开关。此外，请接受本公司1500个型号为CP-765-44电容器的订单。请查阅本电子邮件内所附上的订单明细。可否请您确认44290美元的总价无误？

请务必最晚于11月4日运送至下列地址：

台湾台北　106
忠孝东路3段258号
美丽岛电子公司

美丽岛电子公司采购经理
杰克・方　敬上

“附件”：展翼电脑产品公司采购订单

数量	说明	单价	总价
4,500	电源开关（型号JX-980-12）	8.65 美元	38925.00 美元
1,500	电容器（型号CP-765-44）	0.75 美元	1,125.00 美元
小计			40,050.00 美元
营业税 (10%)			4,050.00 美元
运费和手续费			190.00 美元
总价			44,290.00 美元

Vocabulary and Phrases

1. **place an order for...** 下订单购买某物
   例: I'd like to place an order for ten copies of this book.
   （我想下订单购买这本书 10 册。）
2. **switch** [swɪtʃ] *n.* 开关 & *vt.* 开关(电灯、电器等)
   例: Which switch should I press if I want to turn on the lights here?
   （如果我想打开这里的灯，我该按哪个开关？）
   Please switch the lights off when you leave the office.
   （你离开办公室时请关灯。）
3. **In addition,...** 此外，……
   = Moreover,...
   = Furthermore,...
   = As well,...
   例: Mary is talented in music. In addition, she is quite good at math.
   （玛丽有音乐天赋。而且，她数学也很好。）
4. **confirm** [kənˋfɜm] *vt.* 确认；证实
   confirmation [ˌkɑnfəˋmeʃən] *n.* 确认
   confirm a booking / reservation
   确认预订
   例: Please confirm your reservation in writing by March 1.
   （请在 3 月 1 日以前书面确认您的预订。）
   The cigarette smell confirmed what I had suspected—they had held a party here last night.
   （烟味证实了我先前所怀疑的事情——他们昨晚在这儿办了场派对。）
5. **ship** [ʃɪp] *vt.* (用船、飞机、卡车)运送
   shipment [ˋʃɪpmənt] *n.* 运送；运送的货物
   例: The company ships its goods out to New York every month.
   （这家公司每月将货物运到纽约。）
   The goods are ready for shipment.
   （货物已可以装运。）
6. **unit price** 单价
7. **subtotal** [ˋsʌbˌtotḷ] *n.* 小计
8. **handling** [ˋhændḷɪŋ] *n.* 手续费
   a small handling charge
   少许的手续费

Business Writing Exercises

请按括号中的提示将下列句子译成英文。

1. 我想订购 2000 支圆珠笔和 60 个订书机。(ballpoint pens; staplers)
   ____
2. 可否请您确认价格无误后回复我们？(get back to us)
   ____
3. 按先前议定，我认为这个金额有误。
   ____
4. 请将我们所订的货运送到下列地址。
   ____
5. 请务必最晚于 3 月 15 日交货。谢谢！
   ____

Unit 2

Responding to an Order 回应订单

Basic Structure 基本结构

1. 确认买方订单货品型号（model number）及数量（quantity）。
2. 明确指出订购商品预计交货的时间（expected time of delivery）。
3. 感谢对方并期待继续合作。

- 回应买方订单可用“This letter is to confirm your order.”（这封信旨在确认你的订单。）作为开头。
- 确认型号及数量后，接着可用“We are in the process of filling your order now.”（我们正在出货过程中。）说明一切正常。
- 最后结尾可用“Please rest assured that the goods will be shipped to you... + 时间 .”（请放心，货品将于……送到。）

Sentence Patterns 写作句型 19

Thank you for your order for (quantity + item). / We are pleased to accept your order for (quantity + item).
感谢您订购（某数量的商品）。/ 我们很高兴接到您（某数量的商品）的订单。

- Thank you for your order for 12,000 plastic covers.
  感谢您订购 12000 个塑料套。
  * plastic [ˈplæstɪk] *a.* 塑料的
- We are pleased to accept your order for 5,000 units.
  我们很高兴接受您 5000 个的订单。

This letter is to confirm your order. / We are writing to inform you that we have received your order.
本信旨在确认您的订单。/ 我们来信通知您我们已经收到了您的订单。

- This letter is to confirm your order of March 22.
  本信旨在确认您在 3 月 22 日的订单。
- We are writing to inform you that we have received your order regarding aluminum cans.
  我们来信是为了通知您，我们已经收到您的铝罐订单。
  * aluminum [əˈlumɪnəm] *n.* 铝

We are currently processing your order. / We are in the process of filling your order now.
我们目前正在处理您的订单。/ 我们正按照订单出货中。

- We are currently processing your order for 5,500 pairs of glasses.
  我们目前正在处理您 5500 副眼镜的订单。
- We are in the process of filling your order on tables now.
  我们正按照您的桌子订单出货中。

The goods will be shipped to you by / on (date). / You can expect to receive the items by / on (date).
货物将在（某日期）前 / 当天送达。/ 您可望在（某日期）前 / 当天收到商品。

- The goods will be shipped to you by / on April 30.
  货物将在 4 月 30 日前 / 当天送到您手上。
- You can expect to receive the items by / on December 1.
  您可望在 12 月 1 日前 / 当天收到货物。

We appreciate your business. / Your business with us is appreciated.
感谢您的惠顾。/ 谢谢您和我们做生意。

- Thank you. We appreciate your business.
  谢谢您。感谢您的惠顾。
- Your business with us is very much appreciated.
  非常感谢您和我们做生意。

Model E-mail 电子邮件范例 19

To: Dana Franklin, Dependable **Appliances**[1] Ltd.

From: Terry Zeng, Far Eastern Suppliers

Subject: **Pumps**[2] and **knobs**[3]

Dear Ms. Franklin:

Thank you for your order on Feb. 5 for 1,000 pumps and 4,800 knobs. This letter is to confirm your order. We are **in the process**[4] of filling your order now and will have no problem in **supplying**[5] you with the best **quality**[6] pumps and knobs for your appliances. Please **rest assured that**[7] the goods will be shipped to you on February 7. You can expect to receive the items on that day at about 3:30 p.m. We **appreciate**[8] your business. If you have any questions about your order, please don't hesitate to contact me through e-mail or by phone.

Best regards,

Terry Zeng
Far Eastern Suppliers

中译

收件人：可靠电器股份有限公司，达娜·富兰克林
发件人：远东供应商，泰瑞·曾
主　题：抽水机和旋钮

亲爱的富兰克林小姐：

感谢您在 2 月 5 日订购了 1000 台抽水机和 4800 个旋钮。本信旨在确认您的订单。我们现在正按照订单出货中，对于为贵公司的设备提供最优质的抽水机和旋钮将不成问题。请放心，货物将于 2 月 7 日当天运送到贵公司。您可望在当天下午 3 点半左右收到货物。

感谢您的惠顾。若您对订单有任何问题，请不要犹豫通过电子邮件或电话与我联系。

诚挚的祝福

远东供应商
泰瑞·曾

Vocabulary and Phrases

1. **appliance** [ə'plaɪəns] *n.* （家用）电器（用作公司名称时常用复数）
2. **pump** [pʌmp] *n.* 抽水机
3. **knob** [nɑb] *n.* 旋钮
   doorknob ['dɔr,nɑb] *n.* 球形门把
4. **process** ['prɑsɛs] *n.* 过程 & *vt.* 处理，检查（文件）
   be in the process of + V-ing
   在……的过程中
   例：Ted and Linda are in the process of selling their house.
   （泰德和琳达正在出售他们的房子。）
   Your application is currently being processed.
   （您的申请目前正在处理中。）
5. **supply** [sə'plaɪ] *vt.* 提供（使用下列结构）
   supplier [sə'plaɪɚ] *n.* 供货商
   supply sb with sth　提供某人某物
   = provide sb with sth
   例：How can I carry out the plan if you don't supply me with the money and manpower I need?
   （你若不提供我所需的财力及人力，我怎能完成这个计划？）
6. **quality** ['kwɑlətɪ] *n.* 质量 & *a.* 质量高的
   be of high / good / poor quality
   质量高／好／差
   例：Their new products are of very high quality.
   （他们的新产品质量很高。）
   The store specializes in quality furniture.
   （这家店专营高档家具。）
7. **rest assured + that** 从句　请放心……
   例：You may rest assured that we'll do everything we can to find your missing son.
   （请放心我们会竭尽所能寻找您走失的儿子。）
8. **appreciate** [ə'priʃɪ,et] *vt.* 感激；欣赏
   例：I'd appreciate it if you paid in cash.
   （如果你付现金的话我会很感激。）
   I find it hard to appreciate foreign literature in translation.
   （我发现看翻译作品很难欣赏到外国文学的美妙之处。）

Business Writing Exercises

请按括号中的提示将下列句子译成英文。

1. 感谢您订购 100 套厨房用具。（100 sets of kitchenware）

   ____

2. 我们来信旨在确认你们的订单。

   ____

3. 我们目前正按照你们的订单出货中。

   ____

4. 这批货可望最晚于 3 月 15 日送达。（The goods are expected...）

   ____

5. 您的惠顾我们非常感激并且重视。（very much appreciated and valued）

   ____

Unit 3

Giving Bad News about an Order
告知订单坏消息

Basic Structure 基本结构

1. 告知某型号产品订单将被延误。(The goods you ordered will be delayed.)
2. 说明该产品订单延误的原因。(The goods you ordered are currently out of stock.)
3. 再次道歉，并安抚顾客。(Please bear with us regarding the inconvenience this may cause.)

- 这类电子邮件的目的是告知对方关于订单有所延误等的坏消息，因此应立即回信道歉并解释原因，如果可以，给予对方一个确切的交货日期。
- 在邮件的结尾，可以用“Please bear with us — we are doing everything we can to solve the problem.”(请阁下包容，我们正尽一切努力解决这个问题。)之类的句子来向对方致歉，并让对方知道己方会全力处理这个问题。

Sentence Patterns 写作句型 20

We regret to inform you there is a problem with your order. / Unfortunately, we are unable to fill your order as requested.
我们很遗憾地告知您，您的订单出了问题。/ 不幸的是，我们无法按照您的要求出货。

- We regret to inform you there is a problem with your order of LCD screens.
  我们很遗憾地告知您，您的液晶屏幕订单出了问题。
- Unfortunately, we are unable to fill your order as requested at this time.
  不幸的是，我们这次无法按照您的要求出货。

Regretfully, the (item) you requested is / are currently out of stock. / The product you ordered is now in short supply.
遗憾的是，您要求的（某商品）目前缺货。/ 您所订购的产品目前供不应求。

- Regretfully, the metal hinges you requested are currently out of stock.
  遗憾的是，您要买的金属铰链目前缺货。
- I'm sorry to inform you that the product you ordered is now in short supply.
  很遗憾地通知您，您所订购的产品目前供不应求。

We no longer stock the item you have ordered. / I'm afraid the item you ordered has been discontinued.
您先前订购的货品，我们已不再进货。/ 您先前订购的商品恐怕已停产了。

- I'm sorry, but we no longer stock the item you have ordered in our stores.
  很抱歉，您先前订购的货品，本店已不再进货。
- I'm afraid the item has been discontinued for several months.
  该商品恐怕已经停产几个月了。

Due to circumstances beyond our control, your order will be delayed. / Your order has been delayed because of...
由于发生了我们无法控制的情况，您的订单将被延误。/ 您的订单由于……已被延误。

- Due to circumstances beyond our control, your order will be delayed by a few days.
  由于发生了我们无法控制的情况，您的订单将被延误几天。
- Your order has been delayed because of production breakdowns.
  您的订单由于生产故障已被延误。

We are working hard to V / Please bear with us—we are doing everything we can to V
我们正努力…… / 请阁下包容——我们正尽一切努力……

- We are working hard to rectify the situation.
  我们正努力改善现状。
  * rectify [ˋrɛktə͵faɪ] *vt.* 改正
- Please bear with us — we are doing everything we can to solve the problem.
  请阁下包容——我们正尽一切努力解决这个问题。

Model E-mail 电子邮件范例  20

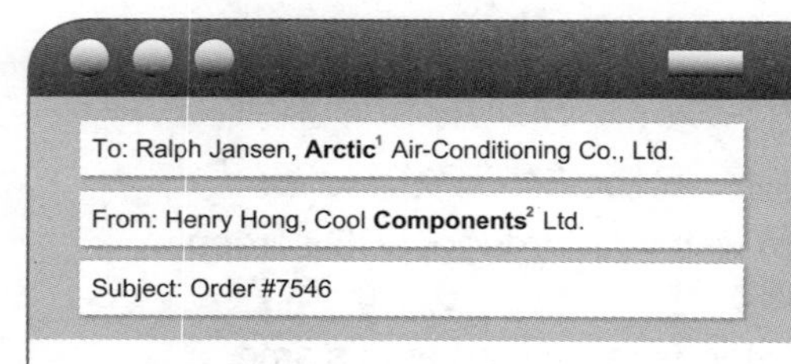

To: Ralph Jansen, **Arctic**[1] Air-Conditioning Co., Ltd.

From: Henry Hong, Cool **Components**[2] Ltd.

Subject: Order #7546

Dear Mr. Jansen:

We **appreciate**[3] receiving your order for **valves**[4] and **drive belts**[5]. Due to **circumstances**[6] beyond our control, your order will be delayed. Some of the valves you ordered (model BG-859) **are** currently **out of stock**[7]. They **are on order**[8] from the manufacturer we **deal with**[9]. We expect to receive them next week.

In addition, unfortunately, we are unable to fill your order exactly as requested because we no longer **stock**[10] the No. 319 valves. I'm afraid that item has been **discontinued**[11]. I don't know any company that is still producing those particular valves.

Sorry for any inconvenience this has caused you. Please **bear with**[12] us regarding the delayed valves—we are doing everything we can to **speed up**[13] the delivery.

Best regards,
Henry Hong
Cool Components Ltd.

中译

收件人：北极空调股份有限公司
拉尔夫・詹森
发件人：冷元件股份有限公司
亨利・洪
主　题：7546 号订单

亲爱的詹森先生：

本公司很感谢收到您订购阀门及传动皮带的订单。由于发生了超出我们控制的情况，您的订单将被延误。您订购的某种阀门（型号 BG-859）目前缺货。本公司已直接向合作厂商订购阀门。预计下星期就能到货。

此外，不幸的是，本公司无法完全按照您的要求出货，因为 319 号阀门本公司已不再进货。该项产品恐怕已经停产。我不清楚是否还有什么公司仍在制造此类阀门。

给您造成任何的不便，本公司深感抱歉。有关阀门的延误，请阁下包容，我们正尽一切努力加快交货。

诚挚的祝福

冷元件股份有限公司
亨利・洪

Vocabulary and Phrases

1. **Arctic** [ˋɑrktɪk] *n.* 北极（与 the 连用，即 the Arctic）
   Antarctic [ænˋtɑrktɪk] *n.* 南极（与 the 连用，即 the Antarctic）
2. **component** [kəmˋponənt] *n.* 零件
   car components　汽车零件
   = car parts
3. **appreciate** [əˋpriʃɪˌet] *vt.* 感激；欣赏

appreciate + N/V-ing　感激……

例：I'd appreciate it if you informed me of the results in advance.
（假如你事先告知我结果的话，我将不胜感激。）

Nobody appreciates being treated like a second-class citizen.
（没人愿意被人当作二等公民对待。）

Kevin's talents were not fully appreci-

ated in that company.
（凯文的才干在那家公司未受充分赏识。）

4. **valve** [vælv] *n.* 气阀，阀门

5. **drive belt** 传动皮带

6. **circumstance** [ˈsɜːkəmˌstæns] *n.* 状况（常用复数）
due to circumstances beyond our control 由于超出掌控的情况
under the circumstances
在这种情况下
例: The meeting has been canceled due to circumstances beyond our control.
（由于发生了超出我们掌控的情况，会议被取消了。）
You did the job well under such circumstances.
（在这种情况下你干得已经不错了。）

7. **be out of stock** 无现货的
be in stock 有现货的

8. **be on order** 在订货中
例: The car parts are still on order.
（这些汽车零件仍在订货中。）

9. **deal with...** 与……做生意
= do business with...
例: Have you ever dealt with that company before?
（你之前和那家公司做过生意吗？）

10. **stock** [stɑk] *vt. & n.* 存货
例: We stock a wide range of office equipment.
（我们有各种办公设备的存货。）
The store carries a large stock of wooden furniture.
（那家店售有很多木制家具。）

11. **discontinue** [ˌdɪskənˈtɪnju] *vt.* 停止，终止（生产）（常用被动语态）
例: Bus route 15 has been discontinued. Please use bus route 22 instead.
（15 路公交车停驶了，请改乘 22 路公交车。）

12. **bear with sb** 容忍某人
例: Brian is under a lot of pressure now. Just bear with him.
（布莱恩现在压力很大。就包容他一点吧。）

13. **speed up...** 加速……
例: The updated system will speed up the registration process.
（更新版的系统将会加快登记流程。）

Business Writing Exercises

请按括号中的提示将下列句子译成英文。

1. 不幸的是，我们无法按照您的要求出货。（fill your order）
____

2. 你们订购的山地自行车（the mountain bikes）目前缺货。（out of stock）
____

3. 很抱歉，你们订购的型号（model）已经停产了。（discontinue）
____

4. 由于发生了我们无法控制的情况，我们必须取消我们的订单。（Due to circumstances beyond our control,...）
____

5. 请阁下包容。我们将会弥补你们的损失。（make up for your losses）
____

Unit 1

Asking for a Payment 要求付款

Basic Structure 基本结构

1. 告知特定型号的产品货款尚未付清。(Your payment for...is overdue.)
2. 说明货款应于某期限前结清。(The payment was due by + 时间 .)
3. 如款项已结清(settle the bill),则请忽略通知(disregard the notice);若持续未支付款项,将采取必要的法律行动(take legal action)。

- 一般对于商场上的付款逾期,可有3种要求付款的方式:
  1. 客气地询问。
  2. 第二种用词较为强硬,会提到先前已通知多次,并坚持对方要在某日之前付款。
  3. 第三种则是会向对方明确表示如再没收到款项,将会采取法律行动。
- 处理催款信永远要谨慎,或许错不在客户,可能款项在寄送过程中遗失或是供应商已收到但忘了记录。

Sentence Patterns 写作句型 21

Our records show your payment for (item) is overdue. / We have not yet received a payment for (item).

本公司记录显示，您订购（某货品）的货款逾期未付。 / 本公司未收到（某货品）的货款。

- Our records show your payment for 1,000 metal tables is overdue.
  本公司记录显示，您订购 1000 张金属桌的货款逾期未付。
- We have not yet received a payment for the Model X8-4 electric switches.
  本公司尚未收到型号 X8-4 电动开关的货款。
  * payment [ˋpemənt] *n.* 款项（可数）；付款（不可数）

We wish to remind you that a payment has not been made on (item). / We request that you settle the invoice concerning (item).

本公司想提醒您，（某货品）的货款尚未支付。 / 本公司请您结清（某货品）的货款。

- We wish to remind you that a payment has not yet been made on your order of drills, model WR-679.
  本公司想提醒您，您订购型号 WR-679 的钻头货款尚未支付。
- We request that you settle the invoice dated April 10 concerning power cables within 30 days.
  本公司请您于 30 天内结清发票日期为 4 月 10 日的电缆货款。
  * settle [ˋsɛtl̩] *vt.* 付清（欠款）

If the payment has already been made, please ignore this notice. / We apologize if you have already settled the bill.

如果您已付款，请不要理会此通知。 / 如果您已付清本账单，本公司深感歉意。

- If payment on the credit card bill has already been made, please ignore this notice.
  如果这份信用卡账单已付清，请不要理会此通知。
- We apologize if you have already settled the bill for the decoration fee.
  如果您已付清装潢费用，本公司致上歉意。

This is the final notice regarding... / This is the last warning concerning...

此为关于……的最后通知。 / 此为关于……的最终警告。

- This is the final notice regarding your overdue account.
  此为关于您逾期账目的最后通知。
- This is the last warning concerning the late payment of your bill.
  此为关于您拖延付款的最后警告。

Unless the payment is made immediately, we will be forced to... / Refusal to submit the payment within (time period) will result in court action.

除非您即刻付款，否则我们将被迫…… / 在（某段时间）内如拒付款项，我们将向法院提起诉讼。

- Unless the payment is made immediately, we will be forced to take legal action.
  除非您即刻付款，否则本公司将被迫采取法律行动。
- Refusal to submit the payment within 45 days will result in court action.
  45 天内如拒付款项，本公司将向法院提起诉讼。

Model E-mail 电子邮件范例 21

To: Gerald Culter, Maroon Industries

From: Eric Zhong, Dragon **Suppliers**[1]

Subject: **Overdue**[2] **account**[3]

Dear Mr. Culter:

Our records show that your payment for 2,500 coils, model #5698-R, is overdue. The shipment was made on October 12, and the payment was **due**[4] by November 30. We kindly request that you **settle**[5] the **invoice**[6] by January 15. If the payment has already been made, please **disregard**[7] this notice.

Please contact us as soon as possible if you have any questions about the payment or this account.

Sincerely,
Eric Zhong
Manager, Accounts Receivable
Dragon Suppliers

To: Accounts Payable Department, Tempo Products

From: Sam Wu, Just In Time Solutions Inc.

Subject: **Outstanding**[8] invoice

Dear Accounts Payable:

Despite **repeated**[9] notices to you on this matter, we have not yet received a payment for the 15,000 rubber **gaskets**[10] that were delivered to you on Feb. 3. This is the final notice regarding this outstanding account. Unfortunately, unless the payment is made immediately, we will **be forced to**[11] **take legal**[13] **action**[12]. If you wish to avoid **court action**[14], please deal with this matter before Sept. 30.

Sam Wu
Assistant Controller
Just In Time Solutions Inc.

中译

收件人：褐红工业　杰拉德・卡尔特
发件人：中国龙供应商　艾瑞克・钟
主　题：账目过期

亲爱的卡尔特先生：

本公司记录显示，您订购 2500 盘型号 5698-R 线圈的货款逾期未付。这批货物于 10 月 12 日装运，而货款应于 11 月 30 日前付清。我们恳请您于 1 月 15 日前结清货款。如果款项已付清，请不必理会此通知。

如果您对此货款或账目有任何疑问，请尽快与我们联系。

中国龙供应商　账款收款部经理
艾瑞克・钟　敬上

中译

收件人：天宝产业　账款部
发件人：及时解决公司　山姆・吴
主　题：未支付的发票

亲爱的账款部：

尽管多次通知您此事，本公司还未收到 2 月 3 日出货给您的 15000 个橡胶垫片的付款。本信为未支付账款的最后通知。遗憾的是，除非款项立即付清，否则本公司将被迫采取法律行动。如果您想避免法院诉讼，请于 9 月 30 日前处理此事。

及时解决公司　助理审计员
山姆・吴

Vocabulary and Phrases

1. **supplier** [sə'plaɪɚ] *n.* 供应商
例: The equipment can be purchased from your local supplier.
（这种设备可向你当地的供应商购买。）

2. **overdue** [ˌovɚ'du] *a.* 过期的；到期未缴的
例: There is an overdue gas bill I need you to pay for.
（这里有一张逾期的煤气账单需要你来支付。）
Linda's baby is two weeks overdue.
（琳达的胎儿已经超过预产期两周了。）

3. **account** [ə'kaʊnt] *n.* 账户
open an account　开户
例: I've opened an account with America First Bank.
（我已在美国第一银行开了户头。）

4. **due** [dju] *a.* 预定的；到期的
例: When is the baby due?
（宝宝预产期是什么时候？）
Payment is due on May 1.
（付款期限为 5 月 1 日。）

5. **settle** ['sɛtḷ] *vt.* 付清（欠款）；结账
例: Please settle your bill before leaving the hotel.
（请您先结账再离开饭店。）

6. **invoice** ['ɪnvɔɪs] *n.* 发票
issue / settle an invoice　开立/结清发票
例: All invoices must be settled by the 25th of every month.
（每月 25 日前应结清所有的发票。）

7. **disregard** [ˌdɪsrɪ'gɑrd] *vt.* 不理会
例: The manager completely disregarded my recommendations.
（经理完全无视我的建议。）

8. **outstanding** [ˌaʊt'stændɪŋ] *a.* 未支付的；杰出的
stand out from / among...
在……中表现特别突出
例: Jennifer has outstanding debts of over $ 1 million.
（詹妮弗有超过 100 万美元的债务未支付。）
We had quite a few applicants for the job, but no one stood out from the rest.
（我们有很多应聘者，但没人表现特别突出。）

9. **repeated** [rɪ'pitɪd] *a.* 重复的

10. **gasket** ['gæskɪt] *n.* 垫圈

11. **be forced to V**　被迫……
例: Our company was forced to relocate from New York to Tokyo.
（我们公司被迫从纽约迁到东京。）

12. **take action**　采取行动（action 为不可数名词）
例: We must take immediate action to cope with that problem before it is too late.
（我们必须立刻采取行动处理那个问题以免为时已晚。）
* cope with...　处理……
= deal with...
= handle...

13. **legal** ['ligḷ] *a.* 法律相关的

14. **court action**　法院诉讼

Business Writing Exercises

请按括号中的提示将下列句子译成英文。

1. 我们的记录显示，你的款项逾期很久未付。(long overdue)
____

2. 我们请您于一周内结清发票上所列的货款。(settle the invoice)
____

3. 如果该款项已付，请不必理会此通知。(disregard the notice)
____

4. 如果你有任何疑问，请尽快与我们联系。
____

5. 除非你们提出充分理由，否则我们将被迫采取法律行动。(Unless sound reasons are given,...)
____

Unit 2

Making Payment　付款

Basic Structure　基本结构

1. 告知对方款项已汇出。(The amount of $22,867 has been transferred to your account.)
2. 说明付款时间及款项金额。
3. 要求对方确认收到款项。(Please confirm that you have received payment for...)

- 此类邮件目的是通知对方已付款，请对方确认，因此内容应包括：

1. 款项的金额数目（the amount of money）。
2. 付款方式（the type of payment）。
3. 订购的产品名称（the name of the product）。

- 商业上的交易通常是以支票付款，如款项众多，可以采取银行转账的方式。
- 如要表示信中有附上支票，可以使用 A check for payment of invoice #9874-BV is included.（随函附上发票编号 #9874-BV 的付款支票。）之类的句子表示。

Sentence Patterns 写作句型 22

Please find enclosed payment for (item). / A check for payment of invoice (#invoice number) is included.

请查收随函附上（某货品）的款项。 / 随函附上发票编号（某号码）的付款支票。

- Please find enclosed payment for the rotary dials we ordered.
  我们随函附上所订购电话拨号盘的支付款项，请查收。
  * rotary [ˋrotərɪ] *a.* 旋转的
  dial [daɪl] *n.* （电话）拨号盘
- A check for payment of invoice #9874-BV is included.
  随函附上发票编号 9874-BV 的付款支票。
  * invoice [ˋɪnvɔɪs] *n.* 发票；（发货或服务）费用清单
  enclosed [ɪnˋklozd] *a.* 随函附上的

The amount of (amount) has been transferred to your account. / A bank transfer of (amount) has been made.

（某金额）款项已汇入您的账户。 / （某金额）的银行汇款已完成。

- The amount of $10,478 has been transferred to your company's account.
  金额为 10478 美元的款项已汇入贵公司账户。
- A bank transfer of $5,608.76 has been made to the account number provided to us.
  银行已将 5608.76 美元的款项汇到您提供给我们的账号。

Please confirm that you have received payment of (amount). / We ask that you provide confirmation of the payment of (amount).

请确认您已收到（某金额）的款项。 / 本公司要求您提供（某金额）款项的付款证明。

- Please confirm that you have received payment of $2,675.
  请确认您已收到金额为 2675 美元的款项。
- We ask that you provide confirmation of the payment of $33,056.
  本公司要求您提供金额为 33056 美元的付款证明。
  * confirm [kənˋfɝm] *vt.* 确认
  confirmation [ˏkɑnfəˋmeʃən] *n.* 证实；证明

Model E-mail 电子邮件范例 22

To: Erin Larkin, Account Manager, Kotter Parts Ltd.

From: Fred Lu, **Top-Flight**[1] Technology

Subject: Payment on Invoice #34967

Dear Ms. Larkin:

Regarding[2] the invoice **dated**[3] November 3, the **amount**[4] of $22,867.00 has been **transferred**[5] to your account. Please confirm that you have received payment for the **automotive parts**[6] that we ordered from your company. This payment **constitutes**[7] full payment of our account.

Yours truly,
Fred Lu
Finance Manager
Top-Flight Technology

中译

收件人：科特零件股份有限公司　客户经理爱琳·拉尔金
发件人：一流技术公司　弗雷德·卢
主　题：发票号码 34967 付款事宜

亲爱的拉尔金女士：

关于 11 月 3 日金额为 22867 美元的发票款项已汇入贵公司账户。请确认您已收到本公司向贵公司订购汽车零件所付的款项。这笔款项可被视作本公司对该账目的全额支付。

一流技术公司　财务经理
弗雷德·卢　敬上

Vocabulary and Phrases

1. **top-flight** [ˌtɑpˈflaɪt] *a.* 第一流的
   = top-notch [ˌtɑpˈnɑtʃ]
   notch [nɑtʃ] *n.* 等级
   例: Gordon is one of our top-flight engineers.
   （戈登是本公司顶尖工程师之一。）
   The service quality at this restaurant dropped a notch recently.
   （这家餐厅的服务质量最近降了一级。）
2. **regarding** [rɪˈgɑrdɪŋ] *prep.* 关于
   = with / in regard to...
   = concerning [kənˈsɝnɪŋ]...
   = about...
   例: The interviewee asked several questions regarding the future of the company.
   （那位面试者问了几个关于公司前景的问题。）
3. **date** [det] *vt.* 注明日期
   例: Thanks for your letter dated March 24.
   （感谢您 3 月 24 日的来函。）
4. **amount** [əˈmaʊnt] *n.* 金额
   例: You will receive a bill for the full amount.
   （您将收到一张全部金额的账单。）
5. **transfer** [trænsˈfɝ] *vt. & vi.* (使)转移
   transfer sth from A to B
   使某物从 A 转往 B
   transfer A to B　把 A 转往 B
   例: How can I transfer money from my account to his?
   （我该如何把钱从我的账户转给他？）
   The patient was transferred to another hospital for better medical treatment.
   （那位病患被转往另一家医院以接受更好的治疗。）
   I transferred to UC Berkeley after my freshman year.
   （我大一读完后转学到加州伯克利大学。）
6. **automotive parts**　汽车零件
   = auto parts
7. **constitute** [ˈkɑnstəˌtjut] *vt.* 被视为；构成
   = make up
   例: Such action was interpreted as constituting a threat to the society.
   （这样的行为被看作是对社会的一种威胁。）
   Asian Americans constitute 25% of the student body at this college.
   （亚裔美国人占这所大学学生总人数的25%。）
   * student body　（学校的）全体学生

Business Writing Exercises

请按括号中的提示将下列句子译成英文。

1. 随函附上一张面额 2500 美元的支票，请查收。(Please find enclosed...)
   ____
2. 金额为 480 欧元的款项已汇入您的账户。(The amount of 480 euros...)
   ____
3. 请确认你们已经接到我方昨日下的订单。(Please confirm that...)
   ____
4. 大量的信息可在网上获得。(A large amount of...)
   ____
5. 大批的旅客目前正在等候火车。(A large number of...)
   ____

Unit 3

Acknowledging Payment 告知收到付款

Basic Structure 基本结构

1. 告知收到某货品的款项，确认对方付款时间及金额。(This letter is to acknowledge receipt of your payment for...)
2. 感谢对方及时付款。(We want to thank you very much for settling the account in such a timely fashion.)
3. 表示期待未来能与对方继续做生意。(We look forward to doing business with you in the future.)

- 这类电子邮件的目的是让对方知道已收到订单款项，特别注意务必将收到款项的金额再说明一次。
- 经常向同一供应商下单的买方，为了避免每次交易付款的不方便，会要求采用记账（open account）方式。在这类交易条件下，买方可在每月、每季或在双方约定的一定期间内付款。

Sentence Patterns 写作句型 23

This letter is to acknowledge receipt of your payment for (item). / Thank you for your payment dated (date).

这封信旨在告知本公司已收到贵公司购买（某货品）的款项。/ 感谢贵公司于（某日期）付款。

- This letter is to acknowledge receipt of your payment for 1,600 metal containers.
  这封信旨在告知本公司已收到贵公司购买 1600 个金属容器的款项。
  * acknowledge [əkˋnɑlɪdʒ] *vt.* 告知收到；确认
- Thank you for your payment dated July 22.
  感谢贵公司于 7 月 22 日付款。

We have received notification of the transference of (amount). / This is to confirm we have received payment in the form of...

本公司已收到（某金额）的转账通知。/ 这封信是要确认本公司已收到……形式的付款。

- We have received notification of the transference of $18,409.
  本公司已收到金额为 18409 美元的转账通知。
  * notification [ˌnotəfəˋkeʃən] *n.* 通知
  transference [trænsˋfɝrəns] *n.* 转移
- This is to confirm we have received payment in the form of a money order.
  这封信是要确认本公司已收到汇票形式的付款。
  * a money order 汇票

The bank record shows our account has been credited with (amount) for payment of your order. / We appreciate the payment of (amount) on your credit account.

银行记录显示贵公司已将订购款项（某金额）汇入本公司账户。/ 本公司感激贵公司使用信用账户付了（某款项）。

- The bank record shows our account has been credited with $5,700 for payment of your order.
  银行记录显示贵公司已将订购款项 5700 美元汇入本公司账户。
- We appreciate the payment of $35,000 on your credit account.
  本公司感激贵公司使用信用账户付了款项 35000 美元。
  * credit account 信用账户（为公司与公司间使用赊贷购买的付款机制）

Model E-mail 电子邮件范例 23

To: Fred Lu, Top-Flight Technology

From: Erin Larkin

Subject: Re: Payment on Invoice #34967

Dear Mr. Lu:

This letter is to **acknowledge**[1] **receipt**[2] of your payment for your order of auto parts. Thank you for your payment dated Dec. 1. We have received **notification**[3] of the transference of $22,867 from your bank into our account. We want to thank you very much for **settling**[4] the account in such a timely **fashion**[5]. In addition, we look forward to doing business with you in the future.

Kind regards,
Erin Larkin

中译

收件人：一流技术公司　弗雷德・卢
发件人：爱琳・拉尔金
主　题：回复：发票号码 34967 付款事宜

亲爱的卢先生：

　　这封信是要告知已收到贵公司订购汽车零件的款项。感谢贵公司在 12 月 1 日付款。本公司已收到由你方银行转至本公司账户 22867 美元的转账通知。本公司十分感谢贵公司能够及时付清账款。此外，本公司期待能在未来与贵公司继续做生意。

　　谨致问候

爱琳・拉尔金

Vocabulary and Phrases

1. **acknowledge** [ək'nɑlɪdʒ] *vt.* 告知收到；确认

   例: Please acknowledge receipt of this e-mail.
   （收到本电子邮件后请告知。）

2. **receipt** [rɪ'sit] *n.* 接收（不可数）；收据（可数）

   on / upon receipt of... 一收到……

   例: The goods will be delivered upon receipt of an order form.
   （订单一到即发货。）

   Make sure you are given a receipt for everything you buy.
   （你买每样东西务必要拿到收据。）

3. **notification** [ˌnotəfə'keʃən] *n.* 通知（不可数）

   notify ['notəˌfaɪ] *vt.* 通知

   advance / prior notification 预先通知

   notify sb of sth
   = inform sb of sth
   告知某人某事

   例: You'll receive notification of our final decision before the next meeting.
   （你会在下次会议前收到我们最终决定的通知。）

   The police must be notified of the date of the demonstration.
   （警方必须被告知游行示威的日期。）

4. **settle** ['sɛtl̩] *vt.* 付清（欠款）

   例: Payment of your account is now overdue, and we must ask you to settle without further delay.
   （您的账目付款已逾期，本公司必须要求您即刻付清。）

5. **fashion** ['fæʃən] *n.* 方式（= manner）

   in a timely / orderly fashion
   以及时/按次序的方式

   例: The fire alarm has gone off. Please evacuate in an orderly fashion.
   （着火警报已经响起。请按次序疏散。）

   * go off （警报器）突然发出巨响
   * evacuate [ɪ'vækjuˌet] *vi. & vt.* （使）疏散

Business Writing Exercises

请按括号中的提示将下列句子译成英文。

1. 本信旨在告知已收到你们购买 30 台台式电脑的款项。

   ____

2. 感谢你们日期为 12 月 3 日的支票付款。

   ____

3. 我们已收到金额为 6000 美元的转账通知。（the transference of $6,000）

   ____

4. 这封电子邮件是要确认本公司已收到您以信用卡方式支付的款项。
   （in the form of credit card）

   ____

5. 银行记录显示贵公司已将订购款项 3050 美元汇入本公司账户。
   （has been credited with $3,050）

   ____

Unit 1

Making Complaints 投诉

Basic Structure 基本结构

1. 针对特定货物的投诉。（I'm writing to make a complaint about...）
2. 说明投诉内容，如货物不符合要求或有瑕疵。（They are not satisfactory for our needs.）
3. 要求对方尽快处理或说明将如何解决。（I hope this problem can be resolved quickly and efficiently.）

- 写投诉信时即使非常不悦，下笔还是要谨慎，避免使用无礼的用语，因为也许错并不在供应商。
- 投诉信应包括以下内容：
  1. 说明所购买的产品或服务是什么。
  2. 解释出现了什么样的问题。
  3. 希望对方如何弥补错误。
- 必须立即投诉向对方反映，才能方便对方调查错误的原因。

Sentence Patterns 写作句型 24

I'm writing to complain about... / I wish to make a complaint about...
我来信投诉有关（某事）。/ 我想投诉（某事）。

- I'm writing to complain about a shipment of brackets made to us.
  我来信投诉有关贵公司运送给本公司的托架一事。
- I wish to make a complaint about an order we received this morning.
  我想投诉有关我们今早收到的订货一事。

There is a problem with (sth). / The (items) we ordered are not satisfactory.
（某事）出了问题。/ 我们订购的（某些货品）令人不满意。

- There is a problem with the delivery of power cables from your company.
  贵公司送来的电缆出了问题。
- The computer chips (model DN-A798) we ordered are not satisfactory.
  本公司订购的型号 DN-A798 电脑芯片令人不满意。

I regret to inform you that the (items) we received were damaged. / Unfortunately, the (items) that were delivered were defective.
我遗憾地通知您，我们收到的（某些货品）受损了。/ 遗憾的是，送来的（某些货品）有瑕疵。

- I regret to inform you that the batteries we received were damaged.
  我遗憾地通知您，我们收到的电池受损了。
- Unfortunately, the cartridges that were delivered were defective.
  遗憾的是，送来的墨盒有瑕疵。

There appear to be some discrepancies between (A) and (B). / There are several inconsistencies in (sth) that need to be addressed.
（A）与（B）似乎有些出入。/（某物）有几项不一致之处需要解决。

- There appear to be some discrepancies between the invoice and the delivery.
  发票上所写与运来的货品似乎有些出入。
- There are several inconsistencies in the shipment that need to be addressed.
  运抵的货物有几项不一致之处需要解决。

I hope the problem with (sth) can be solved quickly and efficiently. / Please tell us how this situation with (sth) will be resolved.
我希望（某事）的问题能快速且有效地得到解决。/ 请告知我们（某事）这个状况要如何解决。

- I hope the problem with the machinery can be solved quickly and efficiently.
  我希望机器设备的问题能快速且有效地得到解决。
- Please tell us how this situation with the faulty equipment will be resolved.
  请告知我们设备有瑕疵这种情况要如何解决。

Model E-mail 电子邮件范例 24

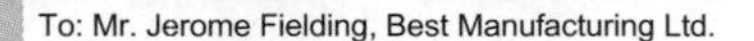

To: Mr. Jerome Fielding, Best Manufacturing Ltd.

From: Paul Feng, **Tall Orders**[1] Inc.

Subject: Problems with the shipment

Dear Mr. Fielding:

I'm writing to **make a complaint about**[2] a shipment of metal rods that was made to us today. There is a problem with the rods, and I must tell you that they are not **satisfactory**[3] for our needs. Although we **specified**[4] that we require rods that are 16 centimeters in **diameter**[5], the items we received were actually 13 centimeters. I hope the problem with this order can be **resolved**[6] quickly and efficiently.

Regards,
Paul Feng
Tall Orders Inc.

To: Sam Zhu, Superior Electronics

From: Ivan Rees, Made-To-Order Solutions

Subject: Delivery of March 23

Dear Mr. Zhu:

I'm writing to complain about a delivery of LED lights on March 23. Unfortunately, some of the lights that were delivered were **defective**[7] (about 10 percent, in fact). In addition, there appear to be a number of **discrepancies**[8] between our agreed purchase price and the prices on the invoice. I have attached a list of the **inconsistencies**[9] that I noticed.

Please tell us how this situation with the damaged LED lights and the price discrepancies will be resolved.

Ivan Rees,
Project Manager
Made-To-Order Solutions

中译

收件人：顶尖制造有限公司
杰罗姆・费尔丁先生
发件人：求精公司　保罗・冯
主　题：运送货物的问题

亲爱的费尔丁先生：

我来信投诉有关今天运抵本公司的金属棒。那些金属棒出了问题，我得告诉你那些货物未能满足本公司的需求。虽然本公司已明确说明所需要的金属棒直径为 16 厘米，但收到的货品实际上却是 13 厘米。我希望这批订货的问题能快速且有效率地得到解决。

诚挚的祝福

求精公司
保罗・冯

中译

收件人：精湛电子公司　山姆・朱
发件人：客制化解决方案公司　伊凡・里斯
主　题：3 月 23 日的货运

亲爱的朱先生：

我来信投诉有关 3 月 23 日送到的 LED 灯。遗憾的是，有些运来的 LED 灯有瑕疵（事实上大约占 10%）。另外，双方公司商定的购买价格与发票上的价格似乎有些出入。我已随函附上我注意到不符之处的清单。

请告知本公司损坏的 LED 灯与价格差异的情况将如何解决。

客制化解决方案公司　项目经理
伊凡・里斯

Vocabulary and Phrases

1. **tall order** 离谱的要求
   例: Getting the project done on time will be a tall order.
   (准时完成这个项目简直比登天还难。)
2. **make a complaint about...** 抱怨……
   = complain about...
3. **satisfactory** [ˌsætɪsˈfæktərɪ] *a.* 令人满意的
4. **specify** [ˈspɛsəˌfaɪ] *vt.* 具体说明,详述
   例: The regulation specifies that you have to keep all test content confidential.
   (那条规定具体说明你必须对考试内容完全保密。)
   * confidential [ˌkɑnfəˈdɛnʃəl] *a.* 机密的
5. **diameter** [daɪˈæmətɚ] *n.* 直径
   radius [ˈredɪəs] *n.* 半径;半径范围
   例: The circle is 30 centimeters in diameter.
   = The circle has a diameter of 30 centimeters.
   (这个圆圈的直径是 30 厘米。)
   This wheel has a radius of 60 centimeters.
   = This wheel is 60 centimeters in radius. (罕)(这个轮胎半径是 60 厘米。)
6. **resolve** [rɪˈzɑlv] *vt.* 解决
7. **defective** [dɪˈfɛktɪv] *a.* 有瑕疵的,有缺陷的(= faulty [ˈfɔltɪ])
8. **discrepancy** [dɪˈskrɛpənsɪ] *n.* 差异
   a discrepancy between A and B A 与 B 间的差异
   例: There are some discrepancies between girls' and boys' performance at school.
   (男孩与女孩的在校表现有些差异。)
9. **inconsistency** [ˌɪnkənˈsɪstənsɪ] *n.* 不一致
   例: There was a great inconsistency between what you said and what you did.
   (你的言行极不一致。)

Business Writing Exercises

请按括号中的提示将下列句子译成英文。

1. 我来信投诉有关你们的送货服务。(shipping service)
   ____
2. 你们今天早上送来的货出了问题。
   ____
3. 我很遗憾地通知您,我们收到的光盘(compact discs)受损了。
   ____
4. 此外,还有质量不一的情况需要解决。
   ____
5. 我希望这些问题能够快速而且有效地得到解决。
   ____

Unit 2

Responding to Complaints　回应投诉

Basic Structure　基本结构

1. 感谢对方反映产品瑕疵。（Thank you for bringing the problem to our attention.）
2. 表示歉意并说明原因。（Please accept our sincere apologies regarding this matter.）
3. 承诺补救方案。（Please rest assured that we are doing everything possible to rectify the problem.）

- 此类邮件目的在于处理顾客的投诉，应注意下列两点：
  1. 假设“顾客或许是对的”（The customer may be right.）。
  2. 要立刻回信确认收到了对方的投诉。如果暂时不能确切地回答对方的问题，应向对方解释正在着手调查，稍后会完整回复。
- 回复投诉信时可在结尾处使用“We are sorry for any inconvenience this may have caused you.”（对于可能已对您造成的任何不便之处，本公司表示歉意。）来表达己方的歉意。

Sentence Patterns 写作句型 25

Thank you for bringing the problem with (sth) to our attention. / We appreciate being informed of...

感谢您让本公司注意到（某物）的问题。/本公司感谢您通知我们……

- Thank you for bringing the problem with the products to our attention.
  感谢您让本公司注意到这些产品的问题。
- We appreciate being informed of the incomplete delivery.
  本公司感谢您通知我们未完全交货一事。

Please accept our apologies for... / We apologize for...

请接受本公司对于……的道歉。/本公司对于……感到抱歉。

- Please accept our apologies for the delay in the shipment.
  请接受本公司对于延迟装运的道歉。
- We apologize for providing you with defective items.
  本公司对于给您提供的那些瑕疵品感到抱歉。

  * defective [dɪˋfɛktɪv] *a.* 有瑕疵的

The problem was due to... / The error was caused by...

这个问题源自于……/该错误起因于……

- The problem was due to a miscommunication between staff members.
  这个问题源自于工作人员之间的沟通不良。
- The error was caused by an improper setting on the machinery.
  该错误起因于机器的设定不当。

  * improper [ɪmˋprɑpɚ] *a.* 不当的

Please rest assured we are doing everything possible to rectify the problem concerning.../ We are doing our utmost to correct this error regarding...

请放心，本公司正在尽一切可能更正有关……的问题。/本公司正在尽最大的努力改正有关……的错误。

- Please rest assured we are doing everything possible to rectify the problem concerning this matter.
  请放心，本公司正在尽一切可能更正这件事的相关问题。
- We are doing our utmost to correct this error regarding your order.
  本公司正在尽最大努力更正您订单中的这项错误。

  * utmost [ˋʌtˏmost] *n.* 最大限度

We are sorry for any inconvenience this may have caused. / We hope this problem hasn't resulted in too much inconvenience for you.

对于可能已造成的任何不便之处，本公司表示歉意。/本公司希望这个问题不会给您造成太多不便。

- We are sorry for any inconvenience this may have caused you or your customers.
  对于可能已对您或您的顾客造成的任何不便之处，本公司表示歉意。
- We hope this problem hasn't resulted in too much inconvenience for you at this time.
  本公司希望这个问题这次不会给您造成太多不便。

Model E-mail 电子邮件范例 25

To: Ms. Patricia Ralston, **Apex**[1] Ltd.

From: Gordon Liang, Neptune Supplies

Subject: Re: Complaint

Dear Ms. Ralston:

Thank you for **bringing** the problem with your order of compact discs **to our attention**[2]. We appreciate being informed of the **scratches**[3] on several of the discs. First of all, please accept our sincere **apologies**[4] regarding this matter. We have **looked into**[5] the matter, and apparently the problem was caused by improper **packaging**[6] material that was used. Please **rest assured**[7] we are doing everything possible to **rectify**[8] the problem with this **procedure**[9] so that it does not happen again. Of course, the **faulty**[10] CDs will be **replaced**[11] immediately at no **charge**[12].

We are truly sorry for any inconvenience this problem has caused you. We are also shipping a package of 300 **complimentary**[13] CDs to you in **recognition**[14] of this problem.

Yours truly,
Gordon Liang
Neptune Supplies

中译

收件人：顶点有限公司　帕特里夏・罗尔斯顿女士
发件人：海王星用品　戈登・梁
主　题：回复：投诉

亲爱的罗尔斯顿女士：

感谢您让本公司注意到光盘订单的问题。本公司感谢您通知我们某些光盘上有刮痕。首先，请接受本公司关于此事件的真诚歉意。本公司已调查过此事，这个问题显然是由于我们使用了不当的包装材料。请放心，本公司正在尽一切可能更正此环节出现的问题，以便此事不再发生。当然，有瑕疵的光盘将会立即免费更换。

对此问题造成的任何不便之处，本公司深表歉意。为表示本公司承认此问题，我们同时还将给您寄一包 300 片的免费赠送的光盘。

海王星用品
戈登・梁　敬上

Vocabulary and Phrases

1. **apex** [ˋepɛks] *n.* 顶点(本文用大写 Apex，做专有名词，为公司名称)
2. **bring sth to sb's attention** 使某事受到某人注意
   例: Thanks for bringing this matter to my attention.
   (感谢你告知我这件事。)
3. **scratch** [skrætʃ] *n.* 刮痕 & *vt.* 刮
   例: The car's paintwork is badly scratched.
   (这辆车的烤漆刮损得很严重。)
4. **apology** [əˋpɑlədʒɪ] *n.* 道歉
   make an apology 道歉
   例: I need to make an apology to you because I opened your letter by mistake.
   (我要向你道歉，因为我不小心拆开了你的信。)
5. **look into...** 调查……
   = investigate...
   例: We're looking into the possibility of merging the two departments.
   (我们正在研究合并那两个部门的可能性。)
6. **packaging** [ˋpækɪdʒɪŋ] *n.* 包装(不可数)
7. **rest assured + that 从句** 放心……
   例: Please rest assured that we'll do our utmost to find your son.
   (请放心，我们会尽全力找到你儿子。)
8. **rectify** [ˋrɛktəˌfaɪ] *vt.* 改正
   例: We must take steps to rectify the situation.
   (我们一定要采取措施整顿这个局面。)
9. **procedure** [prəˋsidʒɚ] *n.* 程序；步骤
10. **faulty** [ˋfɔltɪ] *a.* 有缺陷的
11. **replace** [rɪˋples] *vt.* 取代
    replace A with B 用 B 取代 A
    例: The factory replaced most of its workers with robots.
    (这家工厂用机器人取代了大部分工人。)
12. **charge** [tʃɑrdʒ] *n.* 收费
    at no charge 免费
    = free of charge
    = without charge
    = for free
13. **complimentary** [ˌkɑmpləˋmɛntərɪ] *a.* 免费赠送的
14. **recognition** [ˌrɛkəgˋnɪʃən] *n.* 承认
    in recognition of... 以承认……
    a general / growing recognition of...
    普遍承认……/越来越多的人承认……
    例: There is a general recognition of the need for education reform.
    (大家普遍承认需要教育改革。)

Business Writing Exercises

请按括号中的提示将下列句子译成英文。

1. 感谢您通知我们产品方面的瑕疵。

   ____

2. 对于延迟交货，请接受我们最诚挚的道歉。(the delay in delivery)

   ____

3. 这个问题起因于机械故障(mechanical failure)而非人为疏失。(human negligence)

   ____

4. 请放心，我们正在尽一切可能矫正这个问题。

   ____

5. 对于我们可能已经造成的不便，我们表示歉意。

   ____

Unit 1

Writing to Check on the Progress of a Project
写信询问项目的进展

Basic Structure 基本结构

1. 询问某项目的现况。(I'd like to check on the current status of...)
2. 说明该项目的重要性或说明询问原因。
3. 提醒对方给予最新消息，并表示如有问题可以提出，我方会乐意帮忙。
   (Please give me an update soon. Let me know if I can help you in any way.)

- 这类电子邮件的目的是询问对方项目的进行情况，因此内容可包括：
  1. 询问项目现况（the status of the project)。
  2. 说明项目的重要性（the importance of the project)。
- 为了使项目能顺利完成，可以用“Let me know if everything is fine with this project or if I can help you in any way.”（这个项目是否一切顺利，或者有任何我能帮忙的地方，这些都要让我知道。）之类的话来确保项目能如期完成。

Sentence Patterns 写作句型 26

How is (sth) going? / What is the current status of (sth)?
（某事）怎么样了？ /（某事）的现况如何？

- How is the project going?
  那个项目怎么样了？
- What is the current status of the report?
  那份报告的进展如何？

Have you made much progress on (sth)? / How much of (sth) has been completed so far?
你在（某事）上有了很大的进展吗？ /（某事）至今已完成了多少？

- James, have you made much progress on the PowerPoint presentation?
  詹姆斯，关于那份 PowerPoint 简报你有了很大进展吗？
- How much of the order has been completed so far?
  那份订单至今已完成了多少？

I hope everything is going smoothly with (sth). / Let me know if everything is fine with (sth).
我希望（某事）一切都进展顺利。 / 让我知道（某事）是否一切顺利。

- I hope everything is going smoothly with the assignment.
  我希望这项任务一切都进行得很顺利。
- Let me know if everything is fine with the Peterson account.
  让我知道彼得森客户那儿是否一切都很顺利。
  * account [ə'kaunt] *n.* 客户（指机构）

Is (sth) still on schedule? / Will you be able to finish (sth) by the deadline?
（某事）是否仍按照时间表进行？ / 你能否在最后期限以前完成（某事）？

- Greg, is the product launch still on schedule?
  格雷格，这个产品是否还会如期推出？
  * launch [lɔːntʃ] *n. & vt.* （产品的）发表；上市
- Will you be able to finish the repairs by the deadline?
  你能否在最后期限以前完成修补的工作？

I'm a little concerned about (sth). / I'm getting a bit worried about (sth).
我对（某事）有点儿担心。 / 我开始对（某事）有点儿担忧。

- I'm a little concerned about the advertising campaign.
  我对那场广告宣传活动有点儿担心。
- I'm getting a bit worried about the project.
  我开始对这个项目有点儿担忧。

Model E-mail 电子邮件范例 26

To: Morris Pynn

From: Jerry Zheng

Subject: Client **presentation**[1]

Dear Morris:

I'm writing to **ask about the preparations for the client presentation scheduled for**[2] next week. How is the presentation going? Have you made much **progress**[3] on it yet? I hope everything is going **smoothly**[4] because, as you know, this would be a very big client for us. Let me know if everything is fine with this project or if I can help you in any way.

Best regards,
Jerry.

To: Kelly Clark

From: Larry Winkler

Subject: Market research report

Hello Kelly:

I'd just like to **check on**[5] the current status of the market research report. Is the report still on schedule? Will you be able to finish it by the deadline of April 21? **I'm** a little **concerned about**[6] the research report because the deadline is approaching, and I haven't heard anything from you recently. Please give me an **update**[7] soon. Thanks, Kelly.

Yours truly,
Larry

中译

收件人：莫里斯·平
发件人：杰瑞·郑
主　题：客户简报

亲爱的莫里斯：

我来信是为了询问原定下周进行的客户简报的相关筹备情况。简报现在进行得怎样了？你是否已有很大的进展了呢？我希望一切都很顺利，因为你知道，对我们来说这可是很大的客户。该项目是否一切顺利，或者有任何我能帮忙的地方，这些都要让我知道。

诚挚的祝福

杰瑞

中译

收件人：凯莉·克拉克
发件人：拉里·温克勒
主　题：市场研究报告

凯莉，你好：

我只是想了解一下那份市场研究报告的现况如何。那份报告是否仍按时间表进行？你能否在 4 月 21 日的最后期限前完成那份报告呢？我有点担心那份研究报告，因为最后期限就快到了，而我最近还没从你那儿听到任何消息。请尽快告知我最新的情况。谢谢你，凯莉。

拉里　敬上

Vocabulary and Phrases

1. **presentation** [ˌprɛzənˈteʃən] *n.* 口头报告
   give / make a presentation 做口头报告
   例: The manager asked me to make a short presentation about all of our new products.
   （经理要我针对所有的新产品进行一次简短的口头报告。）

2. **...ask about the preparations for the client presentation scheduled for...**
   = ...ask about the preparations for the client presentation (which is) scheduled for...
   * schedule [ˈskɛdʒul] *vt.* 预定
   be scheduled for + 日期　预定在某日期
   be scheduled to V　预定要……
   例: The meeting is scheduled for Monday afternoon.
   （会议安排在星期一下午。）
   The train is scheduled to arrive at 9 p.m., but it's running five minutes late.
   （那班列车预计晚上 9 点抵达，但却比预定时间晚了 5 分钟。）

3. **progress** [ˈprɑgrɛs] *n.* 进展
   make great / good progress
   有很大的进展/进步
   例: The doctor said that the patient was making great progress.
   （医生说那位病人好了很多。）
   I'm not making much progress with my French.
   （我的法语没多大进步。）

4. **smoothly** [ˈsmuðlɪ] *adv.* 顺利地
   smooth [smuð] *a.* 顺利的；平稳的
   例: The management is taking new measures to ensure the smooth running of the business.
   （管理部门正采取新措施以确保公司顺利运营。）
   The pilot managed to make a smooth landing of the plane.
   （飞行员设法让飞机平稳地降落了。）

5. **check on sth**　检查（是否一切正常）
   例: Can you go upstairs and check on the kids?
   （你能否上楼看看孩子们是否一切安好？）

6. **be concerned about...**　担心……
   = be worried about...
   例: We're deeply concerned about this issue.
   （我们对这个问题深感担忧。）

7. **update** [ˈʌpˌdet] *n.* 最新情况报告 & [ˌʌpˈdet] *vt.* 向（某人）提供最新消息
   例: I'll need regular updates on your progress.
   （针对你的进度我将需要定期知道最新情况。）
   We'll update you on this news story throughout the day.
   （针对此新闻报道，我们一整天都会给您提供最新的信息。）

Business Writing Exercises

请按括号中的提示将下列句子译成英文。

1. 我们国内就业市场的现况如何？（the job market）
   ____
2. 你们在合作计划上有重大进展吗？（your plans of cooperation）
   ____
3. 我希望另辟一条生产线的事情进展顺利。（the establishment of another production line）
   ____
4. 你能否在最后期限之前完工？（by the deadline）
   ____
5. 我开始对我们在海外的投资有点担心。（overseas investments）
   ____

Unit 2

Giving Updates on Projects and Assignments
告知项目及任务的最新情况

Basic Structure 基本结构

告知进行顺利

1. 告知目前情况一切顺利。
2. 告知一切按照计划进行中。

告知出现问题

1. 告知遭遇的问题。
2. 说明问题情况及可能造成的后果。
3. 提出解决方法。

- 这类电子邮件的目的是告知对方项目的进行情况，因此内容会依情况的不同而有所差异。如一切顺利，可以用“All is going well with the assignment.”（这项任务各方面都进展得很顺利。）或“We haven't encountered any major problems.”（我们尚未遭遇任何重大问题）之类的话来表示；假如出现问题，可以用“We've hit a bit of a snag with the specifications.”（关于规格一事，我们遭遇了一点麻烦。）或“Unfortunately, there is a major problem.”（不幸的是，出了重大问题。）之类的句子表示。

Sentence Patterns 写作句型 27

All is going well with (sth). / We haven't encountered any major problems with (sth).
（某事）一切都进行得很顺利。/ 针对（某事）我们尚未遭遇什么重大问题。

- All is going well with the assignment.
  这项任务各方面都进展得很顺利。
- We haven't encountered any major problems with the research.
  该研究我们尚未遭遇什么重大问题。

Everything is going according to schedule regarding (sth). / Things are right on schedule with (sth).
关于（某事）一切都照计划进行中。/（某事）一切正按计划进行中。

- I'm happy to report that everything is going according to schedule regarding the project.
  我很高兴向大家报告，关于该项目一切都照计划进行中。
- Things are right on schedule with the delivery.
  送货正按计划进行。

There's a slight problem with (sth). / We've hit a bit of a snag with (sth).
（某事）出了点小问题。/ 关于（某事），我们遇到了一点麻烦。

- There's a slight problem with the product design.
  产品设计出了点小问题。
- We've hit a bit of a snag with the specifications.
  关于规格一事，我们遇到了一点麻烦。
  * snag [snæg] *n.* （尤指潜在的、意外的、不严重的）麻烦、阻碍
  specifications [ˌspɛsəfəˋkeʃənz] *n.* 规格（常用复数）

Unfortunately, there is a major problem with (sth). / There are some big problems with (sth).
不幸的是，（某事）出了大问题。/（某事）出了些大问题。

- Unfortunately, there is a major problem with one of the components.
  不幸的是，其中一个元件出了大问题。
  * component [kəmˋponənt] *n.* 组成部分；零件；电子元件
- There are some big problems with the supplier.
  那家供应商出了些大问题。
  * supplier [səˋplaɪɚ] *n.* 供应商

Due to (sth), we've fallen behind schedule. / As a result of (sth), we can't meet the deadline.
由于（某事）的缘故，我们的进度落后了。/ 由于（某事）的缘故，我们无法赶上最后期限。

- Due to unforeseen circumstances, we've fallen behind schedule.
  由于不可预见的情况，我们的进度落后了。
  * unforeseen [ˌʌnfɔrˋsin] *a.* 始料不及的
- As a result of equipment malfunctions, we can't meet the deadline.
  由于设备故障的缘故，我们无法赶上最后期限。
  * malfunction [ˌmælˋfʌŋkʃən] *n. & vi.* （机器等）出现故障，失灵

Model E-mail 电子邮件范例 27

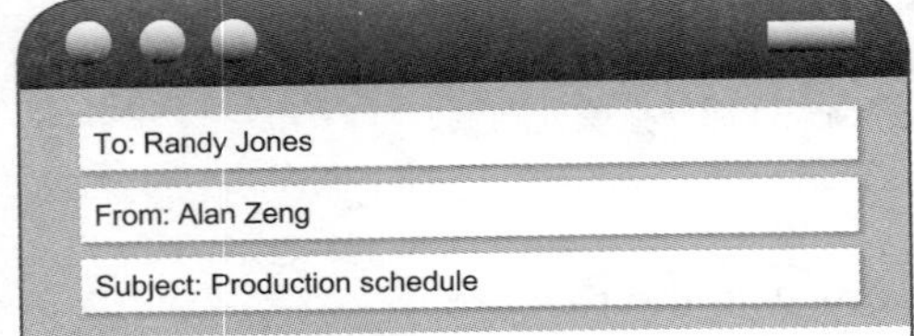

Dear Randy:

Regarding[1] your e-mail about the production **deadline**[2], I want to let you know that all is going well. We haven't **encountered**[3] any **major**[4] problems with the **specifications**[5] or other **requirements**[6]. So, I'm happy to report that everything is going according to schedule regarding production.

Sincerely,
Alan

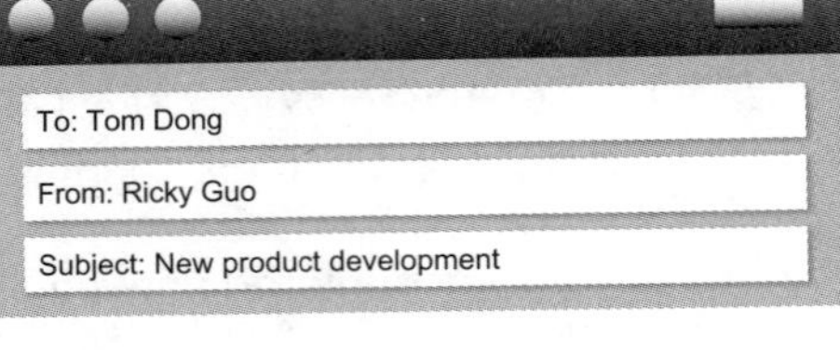

Dear Tom:

I got your e-mail this morning about the **status**[7] of the new product. Unfortunately, there is a major problem with the development. We can't decrease the size of the battery and increase its power **output**[8] at the same time. Due to this **technical**[9] problem, we've **fallen behind**[10] schedule. As a result of this difficulty, we can't meet the deadline. We should arrange a meeting as soon as possible to discuss this problem.

Regards,
Ricky

中译

收件人：兰迪·琼斯
发件人：艾伦·曾
主　题：生产进度表

亲爱的兰迪：

关于你那封生产期限的电子邮件，我想让你知道一切都进行得很顺利。在规格或是其他所需上我们尚未遭遇什么重大问题。所以，我很高兴地向你报告，有关生产一事一切都按照计划进行中。

艾伦　敬上

中译

收件人：汤姆·董
发件人：里奇·郭
主　题：新产品研发

亲爱的汤姆：

我今早收到了你那封有关新产品现况的电子邮件。不幸的是，我们在研发中出现了重大问题。我们无法在缩小电池尺寸的同时，又提高其功率输出。由于这项技术问题，我们的进度已经落后。也由于这个困难，我们无法赶上最后的期限。我们应尽快安排会面讨论这个问题。

诚挚的祝福

里奇

Vocabulary and Phrases

1. **regarding** [rɪˋgɑrdɪŋ] *prep.* 关于
= concerning [kənˋsɝnɪŋ]
= about

2. **deadline** [ˋdɛdˏlaɪn] *n.* 截稿期限；最后期限

3. **encounter** [ɪnˋkaʊntɚ] *vt.* 遭遇

例：We encountered a number of difficulties in the initial stages of the project.
（我们在这个项目的初始阶段遇到了一些困难。）

4. **major** [ˋmedʒɚ] *a.* 主要的
minor [ˋmaɪnɚ] *a.* 次要的

例：There were calls for major changes in the education system.
（有人要求对教育制度进行重大改革。）

5. **specifications** [ˏspɛsəfəˋkeʃənz] *n.* 规格（常用复数）

例：The product was produced exactly to our specifications.
（这个产品完全是按照我们要求的规格生产的。）

6. **requirement** [rɪˋkwaɪrmənt] *n.*
所需的东西或条件（常用复数）
meet sb's requirements
满足某人的需求

例：I'm sure this software package will meet your requirements.
（我确信这个软件套装会满足你的需要。）

7. **status** [ˋstetəs/ˋstætəs] *n.*（进展的）状况，情形

例：What is the current status of our application for funds?
（我们申请资金一事目前进展状况如何？）

8. **output** [ˋaʊtˏpʊt] *n.* 输出（功率）
an output of 100 watts　输出功率 100 瓦

9. **technical** [ˋtɛknɪkḷ] *a.* 技术的

10. **fall behind (...)**　（……）落后
fall behind schedule　进度落后
fall behind with sth　没有及时做某事

例：Study hard, or you will fall behind the other students.
（努力学习，否则你就会落后于其他同学了。）
Tom was ill for three weeks and fell behind with his schoolwork.
（汤姆生病了 3 个星期，因而没有按时做学校作业。）

Business Writing Exercises

请按括号中的提示将下列句子译成英文。

1. 到目前为止，我们尚未遇到任何有关我们设备的重大问题。
____
2. 请放心，一切都按照计划进行中。
____
3. 不过，我们的空调系统出了一点小问题。(air conditioning)
____
4. 由于一次大罢工，我们的进度落后了。(major strike)
____
5. 因此，恐怕我们无法赶上你们的最后期限。
____

Unit 1

Making a Proposal 提议

Basic Structure 基本结构

1. 表明针对某事要进行提议。
2. 详述提议内容。
3. 说明提议优点以说服他人。

- 这类电子邮件的目的是说服收件人做出具体决策，因此内容可包括：
  1. 检视未来所发生之事（future event）。
  2. 表达意见（opinion），并举出客观的事实（facts）来佐证。
- "I have an idea regarding..."（关于……我有个想法……）"My suggestion is that..."（我的建议是……）"What I propose is that..."都可以用来当作发表意见时的开头。

Sentence Patterns 写作句型 28

I propose (that)... / What I propose is...
我提议…… / 我要提议的是……

- I propose (that) we open a new store in the downtown area.
  我提议我们在市中心开家新店。
- What I propose is reducing costs by using contract employees.
  我要提议的是通过使用合同工来降低成本。
  * a contract employee 合同工

I have an idea... / My suggestion is to...
我有个想法…… / 我的建议是……

- I have an idea that might help the company increase profits.
  我有个想法，这可能会帮助公司增加利润。
- My suggestion is to hold the product launch at a five-star hotel.
  我的建议是在一家五星级饭店举办产品发布会。

I'd like to make a point... / Allow me to point out that...
我想要提出一点…… / 容我指出……

- I'd like to make a point regarding the budget preparations.
  关于预算准备我想要提出一点。
- Allow me to point out that our competitors have lowered their prices.
  容我指出我们的竞争对手已降低了价格。

In my opinion, ... / From my point of view, ...
依我之见，…… / 从我的观点来看，……

- In my opinion, we should close our office in France.
  依我之见，我们应该要关闭在法国的办事处。
- From my point of view, the company needs to hire more staff.
  从我的观点来看，公司需要雇用更多职员。

The way I see it, ... / As far as I'm concerned, ...
依我看来，…… / 就我而言，……

- The way I see it, the design is outstanding.
  依我看来，这个设计相当优秀。
- As far as I'm concerned, our product line needs updating.
  就我而言，我们的产品线需要更新。

Model E-mail 电子邮件范例 28

To: Rachel Caruthers

From: Ben Huang

Subject: **Proposal**[1] for company **retreat**[2]

Dear Rachel:

I have an idea regarding the company retreat for next year. What I propose is holding the event at the Mountain Hot Springs Resort. My suggestion is that we **put on**[3] a two-day retreat at this beautiful hotel in a very scenic location. Allow me to **point out**[4] that with so many employees, we could easily get a good discount on rooms at the hotel. **In my opinion**[5], this **venue**[6] would provide a wonderful experience for the staff, and **undoubtedly**[7] could help **boost**[8] **morale**[9]. The way I see it, even though it might be more expensive than what we've paid in past years, the **investment**[10] in the workers would be quite **worthwhile**[11]!

Sincerely,
Ben

中译

收件人：雷切尔・凯鲁瑟斯
发件人：本・黄
主　题：公司拓展休闲游提议

亲爱的雷切尔：

有关明年公司的拓展休闲游，我有一个想法。我要提议的是这次活动在高山温泉度假村举办。我的建议是我们在处于风景优美之地的这家漂亮饭店举办为期两天的拓展休闲游。容我指出，本公司员工众多，因此我们很容易获得饭店房间的折扣。依我之见，这个场地会让员工有很棒的体验，而且无疑也能帮助鼓舞士气。依我看来，虽然可能会比我们过去几年的花费要高，但这项对员工的投资会相当值得！

本　敬上

Vocabulary and Phrases

1. **proposal** [prəˋpozḷ] *n.* 提议
   propose [prəˋpoz] *vt.* 提议，建议
   = suggest
   propose + N/V-ing　建议……
   propose sth to sb　向某人建议某事
   例: The committee put forward a proposal to increase unemployment benefits.
   （委员会提出要增加失业救济金的建议。）
   * benefits [ˋbɛnəfɪts] *n.* （失业）救济金
   John proposed changing the company's logo at the meeting.
   （约翰在会议上建议更换公司商标。）
   Peter proposed a feasible solution to me.
   （彼得给我提了一个可行的解决方案。）
   * feasible [ˋfizəbḷ] *a.* 可行的

Vocabulary and Phrases

2. **retreat** [rɪˋtrit] *n.* (公司)拓展休闲游(即在某度假中心度假，并举办研讨会或娱乐活动)

3. **put on sth** 举办(活动)

   例: The band has decided to put on a concert next month.
   (该乐队已决定在下个月举办一场音乐会。)

4. **point out...** 指出……

   例: My father pointed out to me the dangers of swimming alone.
   (我父亲向我指出独自游泳的危险性。)

5. **In my opinion,...** 依我之见，……

   例: <u>In my opinion</u>, Jane is extremely well qualified for the job.
   = <u>In my view</u>, Jane is extremely well qualified for the job.
   = <u>I'm of the opinion that</u> Jane is extremely well qualified for the job.
   (依我之见，简极其胜任这份工作。)

6. **venue** [ˋvɛnju] *n.* 举行地点(如：音乐厅、体育场、会场)

   例: Please note the change of venue for this event.
   (请注意这场活动的举办地点变了。)

7. **undoubtedly** [ʌnˋdaʊtɪdlɪ] *adv.* 无疑地
   = without (a) doubt

   例: This cake is undoubtedly the best you've ever baked, Mom.
   (妈，这块蛋糕无疑是你烤过的最棒的蛋糕。)

8. **boost** [bust] *vt.* 使增长 & *n.* 帮助；激励

   例: Getting that job did a lot to boost my brother's ego.
   (得到那份工作使我弟弟的自信心倍增。)

   * ego [ˋigo] *n.* 自我；自尊心

   The new measures will give a much needed boost to the economy.
   (新的措施将给经济带来迫切需要的推动力。)

9. **morale** [məˋræl] *n.* 士气

   例: Winning that game was a wonderful boost for the team's morale.
   (赢得那场比赛使球队的士气大振。)

10. **investment** [ɪnˋvɛstmənt] *n.* 投资(与介词 in 连用)
    make an investment in... 投资在……

    例: The company made a large investment in the new technology.
    (该公司在那项新技术上投资了大笔资金。)

11. **worthwhile** [ˏwɝθˋwaɪl] *a.* 值得的
    It's worthwhile to V 做……是值得的

    例: I thought it was worthwhile to clarify that matter.
    (我认为澄清那件事是值得的。)

Business Writing Exercises

请按括号中的提示将下列句子译成英文。

1. 我要提议的是找一位名人来为我们的新产品代言。(endorse our new product)
   ____
2. 容我指出，要做出重大决定的时间点不对。
   ____
3. 从我的观点来看，我们有些设备需要淘汰或更新。(to be replaced or updated)
   ____
4. 依我看来，这一次衰退将持续得比我们认为的更久。(The way I see it, the recession...)
   ____
5. 就我而言，我认为你的想法不可行。
   ____

Unit 2

Responding to Proposals 回应提议

Basic Structure 基本结构

1. 感谢对方所给予的提议。(I appreciate your suggestions on...)
2. 告知对方对于提议的想法。(We considered your proposal very carefully, and we found that...)
3. 说明会如何处理对方的提议。(Your suggestions will be taken into careful consideration.) (I'm afraid we can't adopt your ideas at this time.)

- 在对他人的提议发表意见时，尽量让批评具有建设性，并以正面语气做结尾。
- 假如赞同对方的提议，可以用“I think your proposal is wonderful; thanks for submitting it.”(我认为你的建议很棒；感谢你提出建议。)来表示同意对方的看法。
- 建设性的批评除了指出缺点之外，还应说明事情可以如何改善，否则对方会认为你只是在批判。

Sentence Patterns 写作句型 29

Thank you for submitting your proposal regarding... / I appreciate your suggestions on...
感谢你提出关于……的建议。/ 我很感谢你对于……的建议。

- Thank you for submitting your proposal regarding the new product.
  感谢你提出关于这个新产品的建议。
- I appreciate your suggestions on how to boost our sales.
  我很感谢你对于如何提高我们的销量所提出的建议。

I'll pass along your ideas to... / Your proposal will be sent to...
我会将你的想法转告…… / 你的建议会被送交……

- I'll pass along your ideas to the management committee.
  我会将你的想法转告管理委员会。
- Your proposal will be sent to the vice-president.
  你的建议会被送交副总裁。

The proposals are currently under consideration / review. / We will consider your proposal very carefully...
那些提议目前正在（被）考虑 / 审查中。/ 我们会非常仔细地考虑你的建议……

- The proposals are currently under consideration by management.
  管理层目前正在考虑这些提议。
- We will consider your proposal very carefully and get back to you soon.
  我们会非常仔细地考虑你的建议并尽快回复你。
  * get back to sb 以后再答复某人

I think your proposal is wonderful. / I agree (wholeheartedly) with your suggestion(s)...
我认为你的建议很棒。/ 我（完全）同意你的建议……

- I think your proposal is wonderful, thanks for submitting it.
  我认为你的建议很棒，感谢你提出建议。
- I agree wholeheartedly with your suggestion regarding the budget.
  我完全同意你对该预算的建议。
  * wholeheartedly [ˌholˋhartɪdlɪ] *adv.* 全心全意地

Unfortunately, I don't agree with your suggestion. / I'm afraid we can't use your ideas.
不幸的是，我不同意你的建议。/ 恐怕我们不能采用你的点子。

- Unfortunately, I don't agree with your suggestion about the promotion.
  不幸的是，我不同意你关于宣传广告的建议。
- I'm afraid we can't use your ideas concerning the new office.
  恐怕我们不能采用你关于新办公室的点子。

Model E-mail 电子邮件范例 29

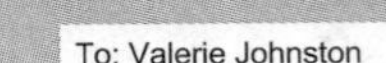

To: Valerie Johnston

From: Charles Thomas

Subject: Advertising campaign

Dear Valerie:

Thank you for **submitting**[1] your proposal regarding the advertising **campaign**[2]. I'll **pass along** your ideas **to**[3] the creative director. Of course, your suggestions will be **taken into consideration**[4] by the director and others involved in the decision process. Personally, I think your proposal is wonderful, although I don't **have the final say**[5] on the matter.

Sincerely,
Charles

To: Valerie Johnston

From: Samantha Guo, Creative Director

Subject: Advertising campaign

Dear Valerie:

I **appreciate**[6] your suggestions on the upcoming advertising campaign. They were creative, interesting and modern. We considered your proposal very carefully; however, I'm afraid we can't **adopt**[7] your ideas at this time. Again, thank you very much for your **input**[8], and I look forward to hearing more from you in the future.

Yours truly,
Samantha

中译

收件人：瓦莱丽·约翰斯顿
发件人：查尔斯·托马斯
主　题：广告宣传活动

亲爱的瓦莱丽：

感谢你提出关于本次广告宣传活动的建议。我会将你的想法转告创意总监。当然，你的建议将由总监及参与决策过程的其他人共同考虑。以我个人来说，我觉得你的提议很棒，不过我在这件事上没有最终决定权。

查尔斯　敬上

中译

收件人：瓦莱丽·约翰斯顿
发件人：创意总监　萨曼莎·郭
主　题：广告宣传活动

亲爱的瓦莱丽：

对你关于即将来临的广告宣传活动所提的建议我深表感激。这些建议颇具创意、有趣且时尚。我们非常认真地考虑了你的建议，但恐怕这次我们不能采用你的点子。再次非常感谢你的投入，而我也期待未来能从你那儿听到更多的建议。

萨曼莎　敬上

Vocabulary and Phrases

1. **submit** [səbˈmɪt] *vt.* 提交；提出

例：The company submitted building plans to the council for approval.
（该公司把建筑计划提交理事会批准。）

2. **campaign** [kæmˈpen] *n.* 运动（为社会、商业或政治目的而进行的活动）

conduct / launch a campaign
领导 / 发起某运动

例：The government launched a campaign against drunk driving.
（政府发起一项禁止醉酒驾车的活动。）

3. **pass along sth to sb**
= pass sth on to sb
转达某事给某人

例：I passed your message on to my manager.
（我把你的留言转给我的经理了。）

4. **take sth into consideration**
考虑某事

例：You should take the applicants' experience and qualifications into consideration when you make the decision.
（你做决定时应该考虑求职者的经验和资格。）

5. **have the last / final say**
有最终决定权

have no say 没有决定权 / 发言权
* say [se] *n.* 决定权 / 发言权

例：The judge has the final say on the sentence.
（法官对该判决有最终决定权。）

The workers had no say in how the company was run.
（那些员工对于公司如何运营没有决定权。）

6. **appreciate** [əˈpriʃɪˌet] *vt.* 感谢

7. **adopt** [əˈdɑpt] *vt.* 采用

例：All three teams adopted different approaches to the problem.
（针对这个问题，3 支队伍采用的方法各不相同。）

8. **input** [ˈɪnˌpʊt] *n.* 投入资源（指时间、知识、思想等）

例：I appreciate your input on this project.
（我感激你在这个项目上的投入。）

Your expert input into our discussions has been very helpful.
（你在我们的讨论中所提供的专业建议相当有帮助。）

Business Writing Exercises

请按括号中的提示将下列句子译成英文。

1. 感谢您针对我们的客户服务（customer service）提出意见（comments）。

____

2. 我会将你的建议转达给我们经理。

____

3. 公司有一些新政策目前正在被评估中。（under evaluation）

____

4. 我确信你的提议将被列入考虑。（...be taken into consideration）

____

5. 首席执行官（CEO）几乎对所有事情有最终决定权。

____

Unit 1

Writing a Report 写报告

Basic Structure 基本结构

1. 简述研究背景。（The survey was conducted over three months... The purpose of the survey was to determine...）
2. 简述研究结果。（We came to the conclusion that...）（Our findings were that...）

- 报告包含的信息是过去已发生的事实，主要目的是提供信息、记录客观事实，使其成为做决定及采取行动的依据。
- 在报告中提供信息时，若做那件事的人不是重点而无须提及时，常使用被动语态表示，如：The inventory work is conducted every six months.（每6个月会进行盘点工作。）
- 通常撰写者会将摘要（summary）、结论（conclusions）和建议（recommendations）放在一起，和报告本身分开，我们称之为执行摘要（executive summary），这样可让大家不用读报告全文就能获得所需信息。

Sentence Patterns 写作句型 30

Passive voice: Often, information in reports is given using the passive voice (i.e. the object is used as the subject).

被动语态：报告中的信息常用被动语态呈现（也就是宾语做主语）。

被动语态（一般现在时 / 一般过去时）

The inventory work is conducted... / The new product was launched...
……进行盘点工作。/ 这项新产品在……发售。

- The inventory work is conducted every six months.
  每6个月会进行盘点工作。
- The new product was launched in the third quarter.
  这个新产品在第三季度发售。

被动语态（现在完成时 / 一般将来时）

Sales have been boosted... / The figures will be calculated...
……销量已被提高。/ ……可以计算出数字。

- Sales have been boosted by the advertising promotion.
  通过广告促销，销量已提高。
- The figures will be calculated using a standard formula.
  使用标准公式就可以计算出这些数字。
  * formula [ˈfɔrmjələ] *n.* 公式

使用 there is / there are / there was / there were 表示"某事物存在"

There are many problems... / There was a delay... ……有许多问题。/……有延迟的情况。

- There are many problems with the current software.
  目前的软件有许多问题。
- There was a delay with the shipment from Indonesia.
  从印尼来的货运有延迟的情况。

In addition, ... / Furthermore, ... 此外，……

- In addition, we need to take a look at our expenditures.
  此外，我们需要看看我们的支出。
- Furthermore, several machines are required to be repaired.
  此外，一些机器需要维修。

What's more, ... / As well, ... 更重要的是，……/同样地，……

- What's more, there will be a new training program implemented soon.
  更重要的是，新的培训方案很快就会实施。
  * implement [ˈɪmpləˌmɛnt] *vt.* 实施，执行
- As well, staff numbers were increased due to our business expansion.
  同样地，由于我们的业务扩展，职员的数量增加了。

Model E-mail 电子邮件范例 30

To: Brent Peterson; Vice-president, marketing

From: Richard Zhang

Subject: Market survey

Dear Brent:

After several months of **extensive**[1] **research**[2], we have completed the market **survey**[3] that you asked my department to **conduct**[4]. I have **attached**[5] a copy of the **executive**[6] **summary**[7] for your **reference**[8].

Best regards,
Richard

中译

收件人：布伦特·彼得森 市场营销副总裁
发件人：理查德·张
主 题：市场调查

亲爱的布伦特：

经过几个月的广泛研究，我们已完成了您要求本部门进行的市场调查。兹已附上执行摘要的附件供您参考。

诚挚的祝福

理查德 敬上

Attachment

Market Survey
Executive Summary*

Background

The survey was conducted over three months, from March 15 to June 20. The purpose of the survey was to **determine**[9] public **reaction**[10] to our new chocolate-flavored soda. In addition, people were asked their opinions of the product's **slogan**[11] and **logo**[12].

The Surveys

In total, 1,200 surveys were completed over the three-month period. Furthermore, 58% (696) of the respondents were female, **while**[13] 42% (504) were male. There were 25 questions on each survey. As well, **participants**[14] were given a free **sample**[15] of the chocolate-flavored soda to try.

* *Note: this is only part of the summary — not the full summary.*

中译

附件

市场调查

执行摘要*

背景

这份调查进行了 3 个多月，时间从 3 月 15 日到 6 月 20 日。本次调查的目的是要判定大众对本公司新巧克力口味的汽水的反应。此外，我们针对该产品的口号和商标询问了民众的意见。

调查

在这 3 个月期间共进行了 1200 份调查。此外，58%（696 位）受访者为女性，而 42%（504 位）受访者为男性。每份调查有 25 个问题。此外，受访者可以得到一瓶免费试喝的巧克力口味汽水。

* 注：这只是摘要的一部分——并非完整摘要。

Vocabulary and Phrases

1. **extensive** [ɪkˈstɛnsɪv] *a.* 广泛的

   例: Extensive research has been done into this rare disease.
   （针对这种罕见疾病已进行了广泛研究。）

2. **research** [ˈrisɜtʃ / rɪˈsɜtʃ] *n.* 研究，调查(不可数)
   carry out / do / conduct research into / on sth 针对某事进行研究

   例: Scientists carried out extensive research into renewable energy sources.
   （科学家对再生能源进行了广泛研究。）

3. **survey** [ˈsɜve] *n.* 民意调查
   carry out / conduct a survey
   进行一项问卷调查

   例: A recent survey showed 65% of those questioned were in favor of the plan.
   （最近的问卷调查显示，有 65% 的受访者支持这项计划。）
   We conducted a survey to find out consumer attitudes towards organic food.
   （我们进行一项问卷调查，旨在了解消费者对有机食品的态度。）

4. **conduct** [kənˈdʌkt] *vt.* 进行
   conduct an experiment / an inquiry / a survey 进行实验 / 询问 / 调查

   例: William is against the practice of conducting experiments on animals.
   （威廉反对在动物身上进行实验的做法。）

5. **attach** [əˈtætʃ] *vt.* 附着
   attachment [əˈtætʃmənt] *n.*
   （用电子邮件发送的）附件
   attach A to B 将 A 加到 B 上面

   例: Attach a small battery to a loudspeaker.
   （把一枚小电池装在扩音器上。）

6. **executive** [ɪgˈzɛkjətɪv] *a.* 执行的

7. **summary** [ˈsʌmərɪ] *n.* 总结，概要
   In summary,... 总的来说，……
   = In brief,...
   = In a word,...
   = In a nutshell,...
   = To sum up,...

   例: In summary, we're not satisfied with your performance at work.
   （总的来说，我们不满意你的工作表现。）

8. **reference** [ˈrɛfrəns] *n.* 参考

for sb's / easy / future reference
供某人/方便/日后参考

例: Keep their price list on file for future reference.
(把他们的价目表建档供日后参考。)
I wrote down the name of the hotel for your reference.
(我写下了那家饭店的名字供你参考。)

9. **determine** [dɪˋtɝmɪn] *vt.* 判定

例: An inquiry was set up by the authorities to determine the cause of the car accident.
(当局已展开调查以判定造成该起车祸的原因。)

10. **reaction** [rɪˋækʃən] *n.* 反应(与介词 to 连用)

例: What's your parents' reaction to the bad news?
(你爸妈对这个坏消息有何反应?)
An emergency fund was set up <u>in reaction to</u> the earthquake.
(已经成立紧急基金以应对这次地震。)

11. **slogan** [ˋslogən] *n.* 口号

12. **logo** [ˋlogo] *n.* 标志

13. **Furthermore, 58% (696) of the respondents were female, <u>while</u> 42% (504) were male.**

* 此处 while 等于 whereas 表示"而",不是"当"。

例: Tom is very good at math, while his brother is absolutely hopeless at it.
(汤姆很擅长数学,而他弟弟绝对是无可救药。)

14. **participant** [pɑrˋtɪsəpənt] *n.* 参与者(与介词 in 连用)

例: Nick was an active participant in the discussion.
(尼克在那场讨论中是个积极的参与者。)

15. **sample** [ˋsæmpl̩] *n.* 样品
a free sample of shampoo
免费试用的洗发乳

Business Writing Exercises

请按括号中的提示将下列句子译成英文。

1. 这些机器每 3 个月维修(maintain)一次。

____

2. 去年超过 50 名员工被裁员。(be laid off)

____

3. 一张新专辑(album)将于 6 月中旬推出。(be released)

____

4. 在我们办公室已经安装好一个电话会议系统。(A teleconferencing system...)

____

5. 目前正在进行盘点的工作。(is being carried out / conducted)

____

Unit 2

Cause and Effect 因果关系

Basic Structure 基本结构

1. 提供所观察到的问题。(From my observations over the past few weeks, I have seen some problems regarding...)
2. 说明该问题的原因。(I realized the problem was caused by...)
3. 提出建议。(Therefore, I highly recommend two things...)

- 此类邮件的目的是在报告中说明某种因果关系，因此可以用 Consequently, ...（因此，……）或者 Thus, ...（所以，……）等用语来表示相互间的因果。
- 结论应简明扼要地指出主要发现，并包含或引导如何改善状况的建议。

Sentence Patterns 写作句型 31

Cause and effect are important and common elements in most reports.
在大部分的报告中，因果关系是相当重要且常见的元素。

...due to + N / As a result of + N, ...
……由于某事。/ 由于某事，……

- The decline in sales was due to more competition in the marketplace.
  销量下降是由于市场上竞争更加激烈所导致。
- As a result of a new management style, our productivity has increased.
  由于新的管理方式，我们的生产力已提升了。

The cause of + N... / be caused by + N
某事的原因…… / ……是由某事所造成。

- The cause of the problem hasn't been determined yet.
  这个问题的原因尚未查明。
- The delays were caused by poor planning.
  规划不善导致了这些延误。

Consequently, ... / As a consequence, ...
因此，……

- The client was not satisfied with our service. Consequently, we lost the deal.
  那位客户并不满意我们的服务。因此，我们失去了这笔交易。
- Our prices are very competitive. As a consequence, sales remain strong.

= Our prices are very competitive. As a result, sales remain strong.
  我们的价格十分具有竞争力。因此，我们的销量还是很好。

...sparked / prompted + N
……引发了某事。

- Lower prices offered by other companies sparked a price war.
  其他公司给出的较低价格引发了价格战。
- The staff cutbacks prompted the union to go on strike.
  这次裁员导致了工会发动罢工。
  * cutback [ˈkʌtˌbæk] *n.* 裁减

Therefore, ... / Thus, ...
因此，…… / 所以，……

- Business has been poor; therefore, we need to consider laying off staff.
  生意一直都不好；因此，我们需要考虑裁员。
  * lay off sb 裁掉某人
- Steve's job performance has been fantastic. Thus, the boss gave him a raise.
  史蒂夫的工作表现一直都非常好。因此，老板给他加薪。

Model E-mail 电子邮件范例 31

To: Ms. Carmen Feng

From: Alanis Fairmont

Subject: **Productivity**[1] & **Efficiency**[2] Report

Attachment: Productivity & Efficiency Report

Dear Ms. Feng:

Here is my report on ways to increase efficiency and productivity in your company. Please find the report included as an attachment to this e-mail. If you have any questions, please don't hesitate to contact me by phone or e-mail.

Best regards,
Alanis

中译

收件人：卡门·冯　女士
发件人：阿拉尼斯·费尔蒙特
主　题：生产力及效率报告
附　件：生产力及效率报告

亲爱的冯女士：

以下是我针对如何提高贵公司效率及生产力的报告。请查阅本电子邮件附件内的报告。如果您有任何问题，请尽管打电话或用电子邮件与我联系。

诚挚的祝福

阿拉尼斯

Attachment

Productivity & Efficiency Report*

From my **observations**[3] **over the past several weeks, I have seen many problems regarding**[4] productivity and efficiency in your company. One problem was due to employee **tardiness**[5]. Many employees **routinely**[6] arrive to work late. The result of this is many hours in lost productivity per month.

Another problem was **inconsistent**[7] quality of the goods **manufactured**[8]. At first, I was **puzzled**[9] by the cause of this problem, but then I realized it was caused by **outdated**[10] **machinery**[11]. As a consequence, this inconsistency will continue to **recur**[12] unless the machinery is **updated**[13]. I believe that, unless new equipment is purchased, this lack of consistent quality will **spark**[14] more customer complaints in the future.

Therefore, I highly recommend two things—**monitoring**[15] staff tardiness more closely and purchasing new machinery.

** Not e: this is only part of the report—not the full report.*

中译

附件

生产力及效率报告*

从我过去几个星期以来的观察中，我已看出很多关于贵公司生产力及效率的问题。其中一个问题是由于员工迟到。许多员工经常迟到。这样的结果就是每个月损失很多小时的生产力。

另一个问题是生产的货物质量不稳定。起初，这个问题的原因令我困惑，但后来我意识到这是老旧机器造成的。因此，除非将机器更新，不然这样不稳定的状况将会反复发生。我认为除非采购新设备，不然在未来这种质量不稳定的情况会引发更多客户投诉。

因此，我强烈建议两件事—— 更密切监控员工迟到的情况及购买新机器。

* 注：这只是报告的一部分——并非完整的报告。

Vocabulary and Phrases

1. **productivity** [ˌprɑdʌkˋtɪvətɪ] *n.* 生产力
2. **efficiency** [ɪˋfɪʃənsɪ] *n.* 效率
3. **observation** [ˌɑbzɚˋveʃən] *n.* 观察
   keep sb under observation
   将某人列入观察
   例: The suspect is being kept under close observation.
   （那位嫌疑犯正被严密监控。）
4. **...over the past several weeks, I have seen many problems regarding...**
   *凡遇上时间副词短语为 "over / during / for + the + past / last + 数字 + 时间名词" 时，主句皆须使用 "现在完成时" 或 "现在完成进行时"。
   例: I've been writing the report during the last two hours.
   （过去两个小时中，我一直在写这份报告。）
5. **tardiness** [ˋtɑrdɪnɪs] *n.* 迟到；缓慢
   tardy [ˋtɑrdɪ] *a.* 迟到的；行动缓慢的
   例: Ted was punished for being tardy for school again.
   （泰德因为上学又迟到而被处罚。）
6. **routinely** [ruˋtinlɪ] *adv.* 常规地，惯常地
   routine [ruˋtin] *n.* 常规
   例: We clean and repair the machines as a matter of routine.
   （我们定期清洗和修理这些机器。）
7. **inconsistent** [ˌɪnkənˋsɪstənt] *a.*
   不一致的
   inconsistency [ˌɪnkənˋsɪstənsɪ] *n.*
   不一致
   例: These findings are inconsistent with those of previous studies.
   （这些调查结果和先前研究的结果不一致。）
8. **manufacture** [ˌmænjəˋfæktʃɚ] *vt.*
   大量生产
9. **puzzle** [ˋpʌzḷ] *vt.* 使困惑
   例: What puzzles me is why Henry quit his stable job.
   （令我不解的是，亨利为什么辞去他那份稳定的工作。）
10. **outdated** [ˌaʊtˋdetɪd] *a.* 过时的
11. **machinery** [məˋʃinərɪ] *n.* 机器（集合名词，不可数）
    a piece of machinery 一台机器
    some machinery 一些机器
    比较:
    a machine 一台机器
    some machines 一些机器

12. **recur** [rɪˋkɝ] *vi.* 再发生

例: The theme of freedom recurs throughout the book.
（关于自由的主题在整本书里一再出现。）

13. **update** [ˌʌpˋdet] *vt.* 使现代化；更新

例: It's about time we updated our software.
（该是我们更新软件的时候了。）

14. **spark** [spɑrk] *vt.* 引发 & *n.* 火花

a shower of sparks 一阵火花

例: Winds brought down power lines, sparking a big fire.
（大风刮断了电线，引起大火。）
The economic reform sparked a storm of protest across the country.
（这项经济改革引发了全国各地的抗议浪潮。）

15. **monitor** [ˋmɑnətɚ] *vt.* 监视

例: Each student's progress will be closely monitored.
（每一位同学的学习情况都受到密切的关注。）

Business Writing Exercises

请按括号中的提示将下列句子译成英文。

1. 大多数人认为物价上涨是因为粮食短缺所致。（food shortages）

____

2. 由于我们的共同努力（joint efforts），我们的销售量增加了不少。

____

3. 王先生基于健康欠佳的原因提前退休了。（due to / because of）

____

4. 他们一直亏钱。因此不得不停业了。（close down their business）

____

5. 美元的疲软导致全球性的争夺黄金。（The weakness of the US dollar sparked...）

____

Unit 3

Connecting Ideas 联结想法

Basic Structure 基本结构

1. 用“相互比较”(comparing)对照两者的差异。
2. 用“举例”(giving examples)印证先前的陈述。
3. 用“强调”(emphasizing)加深印象。
4. 用“改换表述”(rephrasing)做出结论。

- 此类邮件的目的是在报告中说明员工评价结果，因此在报告中可能会出现两种情况：

1. 正面看法。如：Kim is a valuable asset to the company.(金姆对本公司来说是项宝贵的资产。)
2. 负面看法。如：Tony's performance is not up to standard.(托尼的表现未达到标准。)

- 最后结论可用“Upon thorough review of his / her performance this year, I find him / her to be...”(经过对他 / 她今年表现的彻底审查，我发现他 / 她……)

Sentence Patterns 写作句型 32

In good reports, ideas are linked together well. There are several ways to do this, including through comparison, contrast, examples, emphasis, and rephrasing.

在优秀的报告中，所有想法是彼此紧密联结的。有几种方法可以做到这一点，包括通过相互比较、对比、举例、强调及改换表述。

Comparison 比较

Similarly, ... / In the same way, ... 同样地，……

- The first quarter was profitable. Similarly, the second quarter was also good.
  第一季度我们有获利。同样，第二季度的情况也相当不错。
- Advertising can be used to make people aware. In the same way, it can help educate the masses.
  广告可以用来使人们有所了解。同样地，它也能帮助教育大众。

Contrast 对比

However, ... / On the other hand, ... 然而，……/另一方面，……

- We need to speed up production; however, we must not make mistakes.
  我们需要加速生产；然而，我们一定不能犯错。
- On one hand, the outlook for jobs is bleak, but on the other hand, there are still a few hopeful signs.
  一方面，就业市场前景暗淡，但另一方面，还是有一些令人鼓舞的迹象。

  * outlook [ˋaʊtˏlʊk] *n.* 前景
  bleak [blik] *a.* 暗淡无望的

Examples 举例

For example, ... / For instance, ... / To give an example, ...
举例来说，……

- There are several alternatives; for example, we could cut expenses.
  有几种选择；举例来说，我们可以减少开销。

  * alternative [ɔlˋtɝnətɪv] *n.* 可供选择的事物
- We are having shipping problems. To give an example, two deliveries were late last week.
  我们在货运上出了些问题。举例来说，上星期就发生了两次延迟交货的情况。

Emphasis 强调

In fact, ... / As a matter of fact, ... 事实上，……

- Business has improved. In fact, we gained three new major clients this month.
  生意已有所改善。事实上，我们本月就争取到了 3 个大客户。
- The customer is right. As a matter of fact, the customer is always right.
  顾客是对的。事实上，顾客永远是对的。

Rephrasing 改换表述

In other words, ... / To put it another way, ...

换句话说，……

- Changes need to be made. In other words, there will be cutbacks.
  我们需要做出改变。换句话说，我们将会进行裁员。
  * cutback [ˋkʌtˏbæk] *n.* 削减；裁减
- We have some problems. To put it another way, more training is required for staff.
  我们有些问题。换句话说，员工需要更多培训。
  * put [pʊt] *vt.* 说；表达

Model E-mail 电子邮件范例 32

To: Kim Dong

From: Darren Richardson

Subject: Personal **evaluation**[1]

Attachment: Evaluation form

Dear Kim:

Thanks for coming in recently for the annual personal evaluation. Please find my report attached to this e-mail. If you have any questions or comments, feel free to discuss them with me at any time.

Sincerely,
Darren

中译

收件人：金姆·董
发件人：达伦·理查德森
主　题：个人评价
附　件：评价表

亲爱的金姆：

感谢你近日来参与年度个人评价。本电子邮件内附我的报告，请查收。如果你有任何疑问或意见，请随时与我讨论。

达伦　敬上

Attachment

Employee Evaluation

Kim Dong

Since Kim started with the company three years ago, she has **consistently**[2] **earned**[3] good performance **reviews**[4]. Similarly, this year's evaluation is also quite **positive**[5].

After a few years' **employment**[6] with the company, some workers get a bit lazy. However, Kim's performance seems to be improving each year. To give an example, she recently won an international **award**[7] for her design work. As a matter of fact, it was the second year **in a row**[8] that Kim was **recognized**[9] for her **outstanding**[10] talent. Upon **thorough**[11] review of her performance this year, I find Kim to be extremely hardworking and **dedicated**[12]. **To put it another way**[13], Kim is a **valuable**[14] **asset**[15] to the company.

Darren Richardson
Darren Richardson
Creative Director

中译

附件

员工评价

金姆·董

自从金姆 3 年前进公司以来，她在工作上的表现持续获得良好的评价。同样地，今年的评价也是非常肯定的。

一些员工在公司受雇几年后会变得有点懒散。然而，金姆的表现似乎一年比一年好。举例来说，她最近在设计作品上赢得了国际奖项。事实上，这是金姆连续第二年因为杰出的才能而获得认可。在对她今年的表现彻底审查后，我发现金姆非常勤奋且专注。换句话说，金姆对本公司来说是项宝贵的资产。

创意总监

达伦·理查德森

Vocabulary and Phrases

1. **evaluation** [ɪˌvæljʊˋeʃən] *n.* 评价
2. **consistently** [kənˋsɪstəntlɪ] *adv.* 持续地
   例: Jane's work performance has been of a consistently high standard.
   （简的工作表现一直保持高标准。）
3. **earn** [ɝn] *vt.* 赢得
   例: Darren earned a reputation as an expert on tax law.
   （达伦赢得了税法专家的美誉。）
4. **review** [rɪˋvju] *n.* 评价
5. **positive** [ˋpɑzətɪv] *a.* 正面的；肯定的
   negative [ˋnɛgətɪv] *a.* 负面的；否定的
   例: The teacher has a very positive influence on Tony.
   （那位老师对托尼有十分正面的影响。）
6. **employment** [ɪmˋplɔɪmənt] *n.* 雇用
   例: The law prevented the employment of children under ten.
   （法律禁止雇用 10 岁以下的童工。）

7. **award** [əˋwɔrd] *n.* 奖项

8. **in a row** 连续几次地

例: It has been raining for three days in a row.
= It has been raining for three consecutive days.
(已经连下 3 天的雨了。)
* consecutive [kənˋsɛkjutɪv] *a.* 连续不断的

9. **recognize** [ˋrɛkəg͵naɪz] *vt.* 赞赏
recognition [͵rɛkəgˋnɪʃən] *n.* 赞赏；表彰

例: The novel is now recognized as a classic.
(这本小说现在是一部公认的经典名著。)
Paul received an award in recognition of his contribution to the company.
(保罗获奖以表彰他对公司的贡献。)

10. **outstanding** [ˋaʊt͵stændɪŋ] *a.* 出色的

11. **thorough** [ˋθɝo] *a.* 彻底的；全面的

例: Nick has a thorough knowledge of the subject.
(尼克对该主题有全面的了解。)

12. **dedicated** [ˋdɛdə͵ketɪd] *a.* 专心致志的
be dedicated to + N/V-ing
对……专心致志

例: The actress is dedicated to helping the poor.
(那位女演员致力于帮助穷人。)

13. **to put it another way** 换句话说
*put 在此表“说”的意思，常见的句子有：

例: Mike is too trusting; to put it another way, he has no head for business.
(迈克太容易相信人了；换句话说，他没有生意头脑。)
Simply put, we accept their offer or go bankrupt.
= To put it simply, we accept their offer or go bankrupt.
(简单地说，我们要么接受他的条件，不然就会破产。)

14. **valuable** [ˋvæljuəbl̩] *a.* 宝贵的

15. **asset** [ˋæsɛt] *n.* 有价值的人；资产（之后接介词 to）

例: I'm sure Tom will be an asset to the team.
(我肯定汤姆将是该团队的宝贵资产。)

Business Writing Exercises

请按括号中的提示将下列句子译成英文。

1. 同样地，需求量增加时价格会上涨。(prices go up)
____

2. 另一方面，供过于求时价格会下降。(when supply is greater than demand)
____

3. 例如，我们可以回收资源来保护环境。(recycle resources)
____

4. 事实上，我们正在扩大生意。(expand our business)
____

5. 换言之，我们将需要多雇一些员工。
____

Unit 4

Generalizing, Clarifying, Summarizing & Concluding 概括说明、阐明、概述及结论

Basic Structure 基本结构

1. 用“概括说明”（generalizing）点出分析的要点。
2. 用“阐明”（clarifying）详细说明先前的要点。
3. 用“概述”（summarizing）来概括陈述要点以加强语气。
4. 用“结论”（concluding）结尾。

- 阐明事情（clarifying）时，可使用本句“I hope this clears things up.”（我希望这样将事情理清了。）作为结束语。
- 概述（summarizing）的理由很多，包括要把已发生的事情告诉他人，给其作为背景资料；又或是要让人掌握大局状况；也可能是要重申之前的论点有多重要，可使用以下句型：
“In summary,...” “To sum up,...” “In short,...” “In a nutshell,...”

Sentence Patterns 写作句型 33

Generalizing helps to put points into perspective, while clarifying makes things clear. Summarizing provides emphasis and focus, and concluding offers a good ending to a report.

“概括说明”有助于客观判断事情的要点，而“阐明”则可使事情变得清晰。“概述”可提供重点及焦点，而“结论”则可给报告一个很好的结尾。

Generalizing 概括说明

In general, ... / On the whole, ...

一般来说，…… / 总的来说，……

- In general, customers are quite satisfied with our products and service.
  一般来说，顾客相当满意本公司的产品及服务。
- On the whole, the development of the new product has gone smoothly.
  整体而言，新产品的开发进展得很顺利。

Clarifying 阐明

At least... / At any rate, ...

至少…… / 不管怎样，……

- Profits have improved. At least for most products they have.
  利润已有所提高了。至少对他们的大多数产品来说是如此。
- The economy looks better. At any rate, it's better for many industries.
  经济看起来好多了。至少对许多产业来说是如此。

Summarizing 1 概述一

In summary, ... / To sum up, ...

总之，……

- In summary, operational costs have been reduced by 10 percent.
  总之，运营成本已减少了 10%。
- To sum up, we need to deal with clients' complaints satisfactorily.
  总之，我们需要圆满地解决客户投诉。

Summarizing 2 概述二

In short, ... / In a nutshell, ...

简而言之，……

- In short, the incentive programs have improved staff morale.
  简而言之，这项激励方案已经鼓舞了员工的士气。
  * incentive [ɪn`sɛntɪv] *n.* 激励；鼓励
- In a nutshell, the new strategy is a success.
  简而言之，这个新策略是一大成功。
  * nutshell [`nʌt͵ʃɛl] *n.* 坚果壳

Concluding 结论

In conclusion, ... / To conclude, ...
总之，…… / 总之，……

- In conclusion, the situation still requires some monitoring.
  总之，该情势仍需要受到监控。
- To conclude, I recommend a complete overhaul of the program.
  总之，我建议应全面检查该计划。
  * overhaul [ˋovə˞ˏhɔl] *n.* 检修；改造 & [ˏovə˞ˋhɔl] *vt.* 彻底检修；全面检查
  a complete / major overhaul 全面 / 大检查

Model E-mail 电子邮件范例   33

To: Terry Ang

From: Caroline Jones

Subject: **Warehouse**[1] location

Attachment: Summary and recommendation

Dear Ms. Ang:

Here are the results of my **analysis**[2] regarding establishing a new warehouse. I look forward to talking to you further about this during our meeting next week.

Yours truly,
Caroline

中译

收件人：泰瑞 · 昂
发件人：卡罗琳 · 琼斯
主　题：仓库位置
附　件：摘要以及建议

亲爱的昂女士：

以下是本人对于建立新仓库一事所做的分析结果。我期待下星期与你会面时能做更进一步的讨论。

卡罗琳　敬上

Attachment

In general, there are many **benefits**[3] to building a warehouse outside the city. Rental rates are cheaper in the **suburbs**[4]. At least, that is the case in most suburban areas. **Alternatively**[5], purchasing land would **likewise**[6] be much cheaper outside the urban region.

In summary, renting or building a warehouse in a suburban location would be the best option. Obviously, this means that transportation fees would be higher, but this would be more than **offset**[7] by the cheaper **rental**[8] or purchase costs.

In conclusion, I highly recommend **securing**[9] a suburban location for your warehouse as soon as possible to deal with your supply problems. Now is a good time to buy because prices are expected to increase next year.

中译

附件

一般说来，在市区外建立新仓库有很多好处。郊区的租赁费用比较便宜。至少在大多数郊区，情况就是如此。或者，在市区外购买土地也同样便宜得多。

总之，在郊区承租或兴建仓库都是最好的选择。显然，这意味着运输费用会更高，但是这笔费用会被较便宜的租赁或购买费用大大抵销掉。

结论就是，我强烈建议尽快为贵公司仓库找到郊区的位置以解决供应问题。而现在就是绝佳的购买时机，因为土地预计明年就会涨价了。

Vocabulary and Phrases

1. **warehouse** [ˈwɛrˌhaus] *n.* 仓库
2. **analysis** [əˈnæləsɪs] *n.* 分析
   analyze [ˈænlˌaɪz] *vt.* 分析
   例: I was interested in Clark's analysis of the situation.
   （我对克拉克对此状况做的分析感兴趣。）
   The job involves gathering and analyzing data.
   （这项工作包括搜集和分析资料。）
3. **benefit** [ˈbɛnəfɪt] *n.* 益处
   例: The new regulation will be of benefit to management and laborers alike.
   （那条新规章制度将使管理部门和工人同样受益。）
4. **suburb** [ˈsʌbɝb] *n.* 郊区
   suburban [səˈbɝbən] *a.* 郊区的
   in the suburbs　在郊区
   例: I prefer to live in the suburbs rather than live in the city.
   （我比较喜欢住在郊区胜过住在城市里。）
5. **alternatively** [ɔlˈtɝnətɪvlɪ] *adv.* （引出第二种选择或可能的建议）要不，或者
   例: We could go to the Chinese restaurant around the corner, or alternatively, we could try that new Italian restaurant.
   （我们可以去吃街角附近的那家中餐厅，要不，我们可以去试试那家新开的意大利餐厅。）
6. **likewise** [ˈlaɪkˌwaɪz] *adv.* 同样地(= also)
   例: My aunt's second marriage was likewise unhappy.
   （我阿姨的第二次婚姻也不幸福。）

7. **offset** [ˈɔfˌsɛt] *vt.* 抵销(三态同形)
   例: Prices have been raised in order to offset the increased cost of raw materials.
   (为抵销原料成本的增加而提高了价格。)
8. **rental** [ˈrɛntl̩] *n.* 租金; 租赁
9. **secure** [səˈkjʊr] *vt.* (尤指经过努力)获得, 取得
   secure a contract / deal
   订立合同; 达成协议
   例: The team managed to secure a place in the finals.
   (球队取得了参加决赛的一席之地。)

Business Writing Exercises

请按括号中的提示将下列句子译成英文。

1. 总的来说, 大部分乘客对我们的服务感到满意。(On the whole, ...)
   ____
2. 不管怎样, 我们已经扭亏为赢了。(go out of the red)
   ____
3. 简而言之, 我们需要坚持下去才能获得成功。(achieve success)
   ____
4. 简而言之, 你必须评估形势(size up the situation)再做投资。
   ____
5. 总之, 顾客永远是对的。(In conclusion, ...)
   ____

Unit 1

Thanking 感谢

Basic Structure 基本结构

1. 针对对方提供的帮助表达感谢。(Thanks so much for your timely help regarding...)
2. 提出几个具体事项加以赞美。(It was kind of you to assist me in...)(I wish to express my deepest gratitude to you for...)
3. 表达愿意改日回报恩惠。(I'd like toreturn the favor someday soon.)

- 这类电子邮件的目的是要感谢对方。别忽略了感谢他人的重要性，当人觉得自己受到肯定和重视时，他就会表现得更加优秀。
- 不一定要在主题栏中写出感谢的原因，假如原因很长的话，写在邮件正文里比较合适，而主题栏中写下"Thanks"或者"Thank you"就可以了。
- 感谢邮件中应避免提及具体的销售事宜，否则会让对方觉得你的感谢只不过是推销业务的借口。

Sentence Patterns 写作句型 34

感谢他人帮助

Thanks so much for your help regarding... 有关……非常感谢你的帮助。

- Thanks so much for your help regarding organizing the dinner party.
  关于筹办晚宴一事，非常感谢你的帮助。

It was very kind of you to assist me with N / in + V-ing 你人真好能帮我……

- It was very kind of you to assist me in picking out the decorations.
  你人真好能帮我挑选装饰品。
  * pick out sth 精心挑选某物

感谢他人款待

Thank you for a lovely evening / time.
感谢你带给我一个愉快的夜晚/一段美好的时光。

- Thank you for a lovely evening last night.
  感谢你昨晚带给我一个愉快的夜晚。

Your hospitality was greatly appreciated.
非常感谢你的盛情款待。

- Your hospitality on Saturday was greatly appreciated.
  非常感谢你星期六的盛情款待。
  * hospitality [ˌhɑspɪˋtælətɪ] *n.* 好客，款待（不可数）

感谢他人提供服务

It was very thoughtful of you to... 谢谢你……，你想的真周到。

- It was very thoughtful of you to buy a birthday gift for me.
  谢谢你买生日礼物给我，你想得真周到。

I'm very grateful to sb for N / I'm very grateful that sb... 我非常感谢你……

- I'm very grateful that you gave me a ride to the airport last Friday.
  我非常感谢你上星期五开车送我去机场。
  * give sb a ride 让某人搭便车

表示深刻感谢

I wish to express my deepest gratitude...
我想对……表达我最真挚的感谢。

- I wish to express my deepest gratitude for all that you've done for me.
  我想对你为我所做的一切表达最真挚的感谢。

I can't thank you enough for...
我对你的……感激不尽。

- I can't thank you enough for your kindness and thoughtfulness.
  我对你的善良和体贴真是感激不尽。
  * thoughtfulness [ˋθɔtfəlnəs] *n.* 体贴

回应他人感谢信

I was more than happy to V　　我很乐意……

- I was more than happy to help you in this matter.
  我很乐意在这件事上帮助你。

I was more than willing to V　　我很愿意……

- I'm more than willing to help you fix the problem.
  我很愿意帮你解决这个问题。

Model E-mail 电子邮件范例 34

To: Ken

From: Brent

Subject: Thanks

Dear Ken:

Thanks so much for your **timely**[1] help regarding the **arrangements**[2] for my trip to Jakarta recently. It was very kind of you to **assist**[3] me in booking the hotel room and arranging tours of the city. In addition, it was very **thoughtful**[4] of you to pick me up at the airport. I wish to express my deepest gratitude to you for inviting me to meet your family. Your wife's cooking skill is **beyond description**[5]. For me, every moment spent with you will definitely **last**[6] a lifetime. Your hospitality was greatly **appreciated**[7].
In conclusion[8], I can't thank you enough for all that you have done for me, Ken. I hope you can visit Taiwan one day so I can return the **favor**[9].

Yours truly,
Brent

中译

收件人：肯
发件人：布兰特
主　题：感谢

亲爱的肯：

有关你近日安排我前往雅加达的旅程，我非常感谢你及时的帮助。你能帮我预订饭店房间以及安排城市观光真好。此外，你还很体贴地来机场接我。我想对你邀请我去和你家人见面表达最深切的感谢。您夫人的烹饪技巧真是棒到无法形容。对我来说，和你一起度过的每一刻必定会让我终身铭记。你的盛情款待我非常感激。

总之，肯，你为我所做的一切让我感激不尽。我希望你哪天能来台湾玩，好让我可以回报你的恩惠。

布兰特　敬上

To: Brent

From: Ken

Subject: Re: Thanks

Hi Brent:

It was great to get your e-mail. I was more than happy to help you plan your trip to Indonesia. Actually, since I live in Jakarta, it was very convenient for me to book the hotel and arrange the tours. Take care and **keep in touch**[10].

Sincerely,
Ken

中译

收件人：布兰特
发件人：肯
主　题：回复：感谢

嗨，布兰特：

很高兴收到你的电子邮件。我很乐意帮助你规划印度尼西亚之旅。其实，因为我就住在雅加达，对我来说预订饭店和安排观光之行是很方便的。保重，记得要保持联系。

肯　敬上

Vocabulary and Phrases

1. **timely** [ˋtaɪmlɪ] *a.* 及时的
2. **arrangement** [əˋrendʒmənt] *n.* 准备工作(常用复数)
   arrange [əˋrendʒ] *vt. & vi.* 安排
   例: We've already made arrangements for our vacation.
   (我们已经为假期做了安排。)
   Can I arrange an appointment for Monday?
   (我可以安排星期一会面吗？)
   We arranged for a taxi to pick us up at the airport.
   (我们安排一辆出租车到机场接我们。)
3. **assist** [əˋsɪst] *vt.* 帮助；协助
   assist sb in V-ing
   帮助某人从事……
   = help sb (to) V
   例: Ted assisted me in finding a job after I graduated.
   = Ted helped me find a job after I graduated.
   (泰德在我毕业后帮助我找工作。)
4. **thoughtful** [ˋθɔtfəl] *a.* 体贴的
   = considerate
   It is considerate / thoughtful of sb to V
   某人很体贴能……
   例: It was very thoughtful of you to send me the flowers.
   (你人真体贴能送我那些花。)
5. **beyond description** 无法描述
   例: The beauty of the Taroko Gorge is beyond description.
   (太鲁阁峡谷的美真是难以描述。)

6. **last** [læst] *vi.* 持续

例: The meeting lasted (for) one hour and a half.
（那场会议持续了一个半小时。）

7. **appreciate** [əˋpriʃɪˌet] *vt.* 感激

8. **in conclusion** 总之

9. **favor** [ˋfevɚ] *n.* 恩惠

return the favor 回报恩惠

例: Thanks for helping me out. I'll return the favor sometime in the future.
（感谢你帮我。有朝一日我会回报这个恩惠的。）

10. **be / keep in touch (with sb)**
（与某人）保持联系

= be / keep in contact (with sb)

例: Thanks for showing us your products — we'll keep in touch.
（感谢你向我们展示贵公司的产品——我们将保持联系。）

Business Writing Exercises

请按括号中的提示将下列句子译成英文。

1. 你人真好，给我提供这么一个大好机会。（It is kind of you...）
____
2. 我们都非常感谢您的盛情款待。（be greatly appreciated）
____
3. 你真周到，开车送我到机场。（give me a ride）
____
4. 我对你们投入的时间和心力感激不尽。（the time and effort you put in）
____
5. 我非常乐意留下来陪你。（keep you company）
____

Unit 2

Congratulating and Showing Sympathy
祝贺和慰问

Basic Structure 基本结构

祝贺

1. 祝贺对方成功。
2. 对于对方的成功表示高兴。

慰问

1. 慰问对方的不幸。
2. 询问对方是否有可帮忙之处，并表达乐意帮忙之意。

- 这类电子邮件的目的是祝贺或慰问对方。祝贺内容可以简短、正式，也可以是口语而非正式的，视双方关系和情况而定；而慰问信一般而言，内容通常简单诚恳，为了凸显关心之意，可以用手写的形式呈现，真诚地写下想说的话语。
- 祝贺时可用“I'm writing to express my congratulations on...”（我来信是要对……表达我的祝贺。）来开头；而慰问时则可用“I was very sorry to hear about...”（我很遗憾听到关于……）来起头。

Sentence Patterns 写作句型 35

表达祝贺

I'm writing to express my congratulations on... 我来信是对……表达我的祝贺。

- I'm writing to express my congratulations on your promotion.
  我来信是对你升迁一事表示祝贺。

I wish to extend my heartfelt congratulations on... 我衷心祝贺……

- I wish to extend my heartfelt congratulations on your marriage.
  我衷心祝贺你的婚姻幸福。
  * heartfelt [ˋhɑrtˏfɛlt] *a.* 衷心的

表达喜悦或道贺

I was so happy to hear the news about... / I was very glad to hear that...
我很高兴听到关于……的消息。

- I was so happy to hear the news about your newborn baby girl.
  我好高兴听到关于你刚出生的女儿的消息。

Congratulations on... / Congrats on... 恭喜……
  * congrats 为 congratulations 的口语说法。

- Congratulations on your new job!
  恭喜您有了新工作!

表达慰问

I was very sorry to hear about... / I was very sorry to hear that...
我很遗憾听到关于……/ 我很遗憾听到……

- I was very sorry to hear that you were in a car accident recently.
  我很遗憾听到你最近发生车祸的消息。

It was terrible to learn about... / It was terrible to learn that...
获悉关于……的事情真令人难过。/ 获悉……真令人难过。

- It was terrible to learn about the problems you've been having.
  获悉您一直以来所面对的问题，真令人难过。

表达深切的慰问（和死亡有关）

I was deeply saddened to hear about the loss of...
听到关于你失去……的消息让我深感悲伤。

- I was deeply saddened to hear about the loss of your husband.
  听到关于你痛失丈夫的消息令我深感悲伤。

Please accept my condolences on / over...
请接受我对于……的哀悼。

- Please accept my condolences on the death of your son.
  请接受我对于令郎去世的哀悼。
  * condolence [kənˋdoləns] *n.* 哀悼（常用复数）

对病痛或是意外表达慰问

Wishing you a quick / speedy recovery. 祝您早日康复。

- Wishing you a quick recovery, Ben.
  本，祝您早日康复。

Get well soon! 要快点好起来喔!

- Get well soon, Dorothy. All the best.
  多萝西，要快点好起来。祝一切顺利。
  * All the best.（告别用语或书信结束语）祝一切顺利，万事如意。

Model E-mail 电子邮件范例 35

To: Jill

From: Adam

Subject: Congratulations!

Dear Jill:

I'm writing to express my congratulations on your newly-opened business! I was so happy to hear the news that you decided to **fulfill**[1] your dream of **running**[2] your own company. I'm sure you will be very successful.
Again, Jill, congratulations and **all the best**[3] for a successful business!

Best regards,
Adam

中译

收件人：吉尔
发件人：亚当
主　题：恭喜！

亲爱的吉尔：

我来信是要表达对你开业的祝贺！很高兴听到你决定要实现梦想经营自己公司的消息。我相信你会很成功的。

吉尔，再次恭喜并祝事业一切顺利！

诚挚的祝福

亚当

To: Ken

From: Cynthia

Subject: Your traffic accident

Dear Ken:

It was terrible to learn about your serious car accident. I was also very sorry to hear that your wife, Mandy, was **injured**[4] in the accident, too. If you need anything at this time, please don't hesitate to ask me. I'd be more than happy to **help out**[5] in any way I can.
Wishing you and your wife a quick **recovery**[6].

Sincerely,
Cynthia

中译

收件人：肯
发件人：辛西娅
主　题：交通意外

亲爱的肯：

获悉你发生严重车祸的消息真令人难过。我也很遗憾听到你夫人曼蒂也在事故中受伤。如果你此时有任何需要，请尽管告诉我。我会非常乐意尽我所能地提供帮助。

祝你们夫妻早日康复。

辛西娅　敬上

Vocabulary and Phrases

1. **fulfill** [fʊlˈfɪl] *vt.* 实现；履行
   fulfill one's dream / ambition
   实现某人的梦想/抱负
   fulfill a duty / an obligation / a promise
   履行职责/义务/诺言
   例：At the age of 55, Mike fulfilled his ambition of running a marathon.
   （55 岁的时候，迈克实现了他参加马拉松的愿望。）
   Tom failed to fulfill his duties as a father.
   （汤姆未能履行他身为父亲的职责。）
2. **run** [rʌn] *vt.* 经营
   = operate [ˈɑpəˌret]
   run / operate a business　经营企业
   例：Jack has been running his own company since he left school.
   （杰克自从离开学校后就一直在经营自己的公司。）
3. **all the best**　祝一切顺利
   例：I wish you all the best in your college years.
   （祝你在大学期间一切顺利。）
4. **injure** [ˈɪndʒɚ] *vt.* 使受伤
   be badly injured　受重伤
   例：The bus driver was badly injured in the crash.
   （那位公交车司机在撞车事故中受了重伤。）
5. **help (sb) out**　帮助（某人）摆脱困境
   例：When I bought the apartment, my father helped me out with a loan.
   （我买那栋公寓时，我父亲借给我一笔钱救了急。）
6. **recovery** [rɪˈkʌvərɪ] *n.* 恢复（与介词 from 连用）
   recover [rɪˈkʌvɚ] *vi.* 恢复（与介词 from 连用）
   make a full / quick / speedy / slow recovery　完全复原/快速复原/复原很快/复原很慢
   recover from...　从……当中恢复
   例：Henry made a full recovery from the operation.
   （亨利手术后完全康复了。）
   It can take many years to recover from the death of a beloved one.
   （从痛失爱人的痛苦中恢复过来可能要花很多年的时间。）
   * beloved [bɪˈlʌvɪd] *a.* 深爱的

Business Writing Exercises

请按括号中的提示将下列句子译成英文。

1. 我衷心祝福你们的结婚 50 周年纪念。（extend my heartfelt congratulations）
   ____
2. 听说你就要结婚了，我感到非常高兴。
   ____
3. 获悉你们公司已经停业真令人难过。（your company had closed down）
   ____
4. 请接受我对于令尊去世所致的哀悼。（accept my condolences）
   ____
5. 祝你周末愉快！
   ____

Unit 1

Writing a Cover Letter 写求职信

Basic Structure 基本结构

1. 说明所欲应聘的职位名称。(I'm writing to apply for the position of...)
2. 简述自身的经历(qualifications and experience)、特质(qualities)、职业目标(career goals)等。
3. 表示期待与对方见面。(I'd be very pleased to meet with you at your convenience.)

- 基本上求职信是推荐信的一种,也就是要试着去推销自己(sell yourself),较好的方式是用一封求职信(a cover letter)来应聘这个职位,并说明简历(curriculum vitate or résumé)已随信附上。简历上应详列个人背景、教育、取得资格和经历,而这些就不必在短信中重复详述。
- 为了增加求职信的说服力,可以用"I have (number) years of experience in (job area)."(我在(某工作领域)有(多少)年的经验。)和"My qualifications include..."(我的资历包括……)之类的话推销自己。

Sentence Patterns 写作句型 36

引言

I'm writing to apply for... 我来信应聘……

- I'm writing to apply for the position of salesperson with your company.
  我来信应聘贵公司的业务员一职。

I saw your ad for a (job title) advertised in the newspaper / on the website.
我看到了贵公司刊登在报纸 / 网站上的招聘广告。

- I saw your ad for a graphic designer advertised on the website greatjobs.com.
  我看到了贵公司刊登在 greatjobs.com 网站上的征求平面设计师的广告。

介绍经历

I have (number) years of experience in (job area).
我在（某工作领域）有（多少）年的经验。

- I have two years of experience in marketing.
  我有两年市场营销的经验。

My qualifications include... 我的资历包括……

- My qualifications include a master's degree and two professional certificates.
  我的资历包括一个硕士学位和两张专业证书。
  * qualification [ˌkwɑləfəˋkeʃən] *n.* 资历
    certificate [səˋtɪfəkət] *n.* 合格证书

描述个性及才能

I am a very (adjective) individual. 我是一个很……的人。

- I am a very hardworking individual.
  我是一个非常勤奋的人。

My qualities include (noun) and (noun). 我的特质包括了……和……

- My qualities include persistence and dedication.
  我的特质包括了毅力和奉献精神。
  * persistence [pɚˋsɪstəns] *n.* 坚持, 执着
    dedication [ˌdɛdəˋkeʃən] *n.* 奉献

谈论目标

My career goal is to... 我的职业目标是……

- My career goal is to work for a large, multinational company.
  我的职业目标是任职于大型跨国公司。
  * multinational [ˌmʌltɪˋnæʃənəl] *a.* 跨国的

My aim is to... 我的目标是……

- My aim is to grow as a professional and develop my skills to their utmost.
  我的目标是成为专业人士，并将我的技能发挥到极致。
  * utmost [ˋʌtˌmost] *n.* 极限

结束语

I'd be very pleased to meet with you... 我会很高兴和您见面……

- I'd be very pleased to meet with you at your convenience.
  我很乐意在您方便时和您见面。

I look forward to discussing this opportunity... 我期待讨论这个机会……

- I look forward to discussing this opportunity with you further.
  我期待与您更进一步讨论这个机会。

Model E-mail 电子邮件范例 36

To: Keith McGuire / Human Resources Manager, Eastern Airlines

From: Marlene Zhong

Subject: Flight attendant position

Attachment: Résumé

Dear Mr. McGuire:

I'm writing to **apply**[1] for the **position**[2] of **flight attendant**[3] that I saw **advertised**[4] in the *Times Newspaper*. My qualifications include a certificate from the Sky Institute, where I received flight attendant training for one year. I am a very hardworking and **outgoing**[5] **individual**[6]. My other **qualities**[7] include **professionalism**[8] and **resourcefulness**[9]. My career goal is to gain **employment**[10] with a major airline such as Eastern Airlines.
I'd be very pleased to meet with you at your convenience. I look forward to discussing this opportunity with you in the near future.

Yours truly,
Marlene Zhong

中译

收件人：东方航空公司人力资源部经理／凯斯·迈克奎尔
发件人：玛琳·钟
主　题：空中乘务员一职
附　件：个人简历

亲爱的迈克奎尔先生：

我来信是要应聘我在《时代日报》的广告上看到的空中乘务员一职。我的资历包含一份航天学院颁发的证书，我在那儿接受了为期一年的空中乘务员培训。我是个非常努力、外向的人。我的其他特质包括具有专业素养和机智敏捷。我的职业目标是受雇于像东方航空公司这样的大型航空公司。

我很高兴在您方便时与您见面。我期待在不久的将来能与您讨论这个机会。

玛琳·钟　敬上

Vocabulary and Phrases

1. **apply** [əˈplaɪ] *vi.* 申请
   apply for a scholarship / visa / job
   申请奖学金/签证/工作
   apply to + 学校/公司
   向学校/公司提出申请
2. **position** [pəˈzɪʃən] *n.* 职务
   例: Henry has held the position of sales director since 2001.
   (亨利自从 2001 年起就担任销售主管的职位。)
3. **flight attendant** 空中乘务员
4. **advertise** [ˈædvəˌtaɪz] *vt.* 登广告
   advertisement [ˈædvəˌtaɪzmənt] *n.* 广告(简称 ad)
   例: The cruise was advertised as a journey that is worth every penny.
   (这次航行被宣传为一趟值得每分钱的旅程。)
   I put an ad in the local newspaper to sell my used car.
   (我在地方报纸上刊登广告卖我的二手车。)
5. **outgoing** [ˈaʊtˌgoɪŋ] *a.* 外向的
   例: Jack is an outgoing and lively person.
   (杰克是既开朗又活泼的人。)
6. **individual** [ˌɪndəˈvɪdʒuəl] *n.* 个人
7. **quality** [ˈkwɑlətɪ] *n.* (尤指好的)人品;质量
   例: I'm confident that our products are of top quality.
   (我有信心我们的产品都拥有最高质量。)
8. **professionalism** [prəˈfɛʃənəˌlɪzəm] *n.* 专业素养
9. **resourcefulness** [rɪˈsɔrsfəlnɪs] *n.* 机智敏捷
   resourceful [rɪˈsɔrsfəl] *a.* 机敏的
10. **employment** [ɪmˈplɔɪmənt] *n.* 雇用
    conditions / terms of employment
    雇佣条件/条款
    例: College graduates are finding it more and more difficult to gain employment.
    (大学毕业生感觉找工作越来越难。)

Business Writing Exercises

请按括号中的提示将下列句子译成英文。

1. 我来信应聘贵公司的营销经理一职。(the position of marketing manager)
   ____
2. 我有 5 年担任秘书的经验。
   ____
3. 我是一个乐观进取的人。
   ____
4. 我的职业目标(career goal)是成为一位有名的律师。
   ____
5. 我的工作职责包括与老客户保持联系及寻找新客户。(My job duties include...)
   ____

Unit 2

Writing a Follow-up Letter
写后续信函

Basic Structure 基本结构

1. 对面试官先前拨冗进行面试表达感谢，并表示对该工作相当有兴趣。（Thank you for taking the time to talk to me...）
2. 再次强调自己的人格特质。
3. 表示自己会对公司做出重要贡献。（I can make a valuable contribution to your company.）

- 后续信函的目的是对面试官表达感谢，并再次推荐自己是这个职位的最佳人选。
- 为了向对方表示求职的诚心，可以用"I'm quite interested in working for..."（我对于任职于……很感兴趣。）和"I'm very hopeful that..."（我很希望……）之类的话推销自己。
- 因为拼写和标点错误会给他人留下不细心或不在乎的印象，所以要在发送邮件之前仔细检查用字遣词及拼写和标点符号是否正确。

Sentence Patterns 写作句型 37

Thank you for taking the time... 感谢您抽出宝贵的时间……

- Thank you for taking the time to interview me earlier today.
  感谢您抽出宝贵的时间在今天稍早对我进行面试。

I appreciate your inviting me... 我感激您邀请我……

- I appreciate your inviting me to be interviewed at your office.
  我感谢您邀请我去您办公室进行面试。

I'm quite interested in working for... 我对于任职于……很感兴趣。

- I'm quite interested in working for your company as a programmer.
  我对于在贵公司担任程序设计员很感兴趣。

I'm very hopeful that... 我很希望……

- I'm very hopeful that you will select me to fill the position.
  我很希望您会选择我来填补该职位。

As I mentioned during the interview,... 正如我在面试期间提到的，……

- As I mentioned during the interview, I have a master's degree in chemistry.
  正如我在面试时所提到的，我拥有化学硕士学位。

I wish to reiterate that... 我希望重申……

- I wish to reiterate that I am fluent in three languages.
  我希望重申，我精通 3 种语言。
  * reiterate [riˋɪtəˏret] *vt.* 反复地说；重申

The interview has reinforced my conviction that... 这次面试加强了我的信念……

- The interview has reinforced my conviction that I am well suited for this job.
  这次面试坚定了我的信念，就是我相当适合这份工作。
  * reinforce [ˏriɪnˋfɔrs] *vt.* 加强
  conviction [kənˋvɪkʃən] *n.* 坚定的看法（或信念）

After the interview, I am convinced that... 这次面试后，我深信……

- After the interview, I am convinced that this position is a truly great opportunity.
  在这次面试过后，我深信这个职位是个很棒的机会。

Thank you again for considering... 再次感谢您考虑……

- Thank you again for considering me for the job.
  再次感谢您考虑让我接下这份工作。

Once again, I appreciate the opportunity... 我再次感谢能有这个机会……

- Once again, I appreciate the opportunity to compete for this position.
  我再次感谢能有这个机会来竞争这个职位。

Model E-mail 电子邮件范例 37

To: Mr. Richard Branson / General Manager, Tiger Technologies

From: Jenny Wang

Subject: Thank you for the interview

Dear Mr. Branson:

Thank you for taking the time to talk to me yesterday afternoon. I appreciate your inviting me to your office for the interview. I'm quite interested in working for Tiger Technologies, and I'm extremely **hopeful**[1] you will choose me for the job.

As I mentioned during the interview, I am hardworking and very resourceful. I wish to **reiterate**[2] that I am **a fast learner**[3] and don't mind **working overtime**[4] to make sure every **task**[5] is **performed**[6] to the highest **standards**[7].

The interview yesterday has **reinforced**[8] my belief that I am definitely the right person for this job. After the interview, I am **thoroughly**[9] convinced I can make a **valuable**[10] **contribution**[11] to your company.

Thank you again for considering my application.

Yours truly,
Jenny Wang

中译

收件人：理查德・布兰森先生 老虎科技公司总经理
发件人：珍妮・王
主　题：感谢您的面试

亲爱的布兰森先生：

感谢您昨天下午抽空与我面谈。我很感激您能邀请我至贵公司办公室进行面试。我对于为老虎科技工作一事很感兴趣，我非常希望贵公司能选择我做这份工作。

正如我在面试期间所提及，我工作努力且善于随机应变。我想重申，我学习速度很快且不介意加班，目的就是要确保以最高标准完成每项工作。

昨天的面试坚定了我的信念，我绝对会是这份工作的适当人选。面试后，我绝对相信我可以为贵公司做出宝贵的贡献。

再次感谢您将我这次求职申请纳入考虑。

珍妮・王　敬上

Vocabulary and Phrases

1. **hopeful** [ˋhopfəl] *a.* 抱有希望的
   be / feel hopeful about sth
   对某事抱有希望
   例: I feel hopeful that we'll find a suitable candidate for this job very soon.
   （我对很快能找到这份工作的适当候选人抱有希望。）
   Gary is not very hopeful about the outcome of the interview.
   （盖瑞对那场面试的结果不抱太大的希望。）
2. **reiterate** [riˋɪtəˏret] *vt.* 反复地说；重申
   例: The government reiterated its refusal to yield to terrorists.
   （政府重申拒绝向恐怖分子屈服。）
   * yield to... 屈服于……
3. **a fast learner** 快速学习者
   = a quick learner
4. **work overtime** 加班工作
5. **task** [tæsk] *n.* 工作；任务
   perform / carry out / complete a task
   执行/执行/完成任务
   例: We usually ask interviewees to perform a few simple tasks on the computer.
   （我们通常会要求应试者在电脑上执行一些简单的任务。）
   Gaining this information was no easy task.
   （获得这份资料绝不是件简单的事。）
6. **perform** [pɚˋfɔrm] *vt.* 执行
7. **standard** [ˋstændɚd] *n.* （质量的）标准；水平
   例: This piece of work is <u>below standard</u>.
   = This piece of work is <u>not up to standard</u>.
   （这份工作未达标准。）
8. **reinforce** [riˋɪnfɔrs] *vt.* 加强
   例: Nick's rude attitude towards others reinforced my dislike of him.
   （尼克对他人粗鲁的态度让我更不喜欢他了。）
9. **thoroughly** [ˋθɝolɪ] *adv.* 彻底地；全面地
   thorough [ˋθɝo] *a.* 彻底的；全面的
   例: The police carried out a thorough investigation into the murder.
   （警方针对这起谋杀案展开了全面的调查。）
10. **valuable** [ˋvæljəbḷ] *a.* 珍贵的
    例: I'm convinced that I'll be a valuable asset to your company.
    （我深信我对贵公司来说会是珍贵的资产。）
11. **contribution** [ˏkɑntrəˋbjuʃən] *n.* 贡献
    make a contribution to...
    对……做出贡献
    例: To our surprise, Henry made a positive contribution to the success of the project.
    （令我们惊讶的是，亨利对这个项目的成功做出了积极的贡献。）

Business Writing Exercises

请按括号中的提示将下列句子译成英文。

1. 感谢你抽出时间来看我的报告。
   ____
2. 我对担任贵公司销售代表颇有兴趣。（a sales representative）
   ____
3. 我想重申我是一个负责任又有效率的人。（I wish to reiterate that...）
   ____
4. 那次面试坚定了我自主创业的信念。（set up my own business）
   ____
5. 我想再次感谢你的好意邀请。（your kind invitation）
   ____

Unit 1

Making and Asking for an Offer
报价及要求报价

Basic Structure 基本结构

1. 表明已收到目录。(Thank you for sending us your catalog.)
2. 简述己方对对方哪项产品感兴趣，并且提出报价。(We are interested in doing business with you. Regarding..., we're prepared to offer...)
3. 针对某项产品提出低于目录上的报价。(Due to our finances, we are prepared to offer...)

- 这类电子邮件的目的是向对方要求报价，为了让对方知道我方的报价底限，可以用 "Our best offer on this is (price)." (对于这项产品，我们提供最优惠的报价是（某价格）。) 或者 "The best we can ofter is (price)." ((某价格) 是我们能提出最优惠的价格。) 之类的话来告知对方。

Sentence Patterns 写作句型 38

提出报价（一）

We are / I am prepared to offer... 我们/我愿意提供……的报价。

- We are prepared to offer you a price of $3 per unit.
  我们愿意提供给你单价 3 美元的价格。

We / I believe that (price) is fair. 我们/我相信（某价格）是合理的。

- I believe that $100,000 is fair.
  我相信 10 万美元的价格是合理的。

提出报价（二）

How would you feel about...? 你觉得……如何？

- How would you feel about an offer of $350,000?
  你觉得 35 万美元这个报价如何？

Is (price) acceptable to you? （某价格）你可以接受吗？

- Is $1.5 million acceptable to you?
  150 万美元你可以接受吗？

提出固定报价

Our best offer on this is (price). 对于这项产品，我们提供的最优惠的报价是（某价格）。

- Our best offer on this is $75 per chair manufactured.
  对于这项产品，我们提供的最优惠的报价是每把椅子制造费 75 美金。

The best we can offer is (price). （某价格）是我们能提出的最优惠的价格。

- The best we can offer is $20 per item.
  每件物品 20 美元是我们能提出的最优惠的价格。

要求报价（一）

What do you propose regarding...? 关于……，你们有何建议？

- What do you propose regarding the unit price?
  关于这个单价，你们有何建议？

What are you prepared to offer on...? 针对……你们愿意提出什么样的报价？

- What are you prepared to offer on the sofa?
  针对这张沙发，你们愿意提出什么样的报价？

要求报价（二）

What did you have in mind regarding...? 对于……你心里有什么想法？

- What did you have in mind regarding the components?
  对于这些零件，你心里有什么想法？

Could you tell me what your offer is on...? 针对……，你能告诉我你的报价是多少吗？

- Could you tell me what your offer is on the cables?
  针对这些电缆，你能告诉我你的报价是多少吗？

Model E-mail 电子邮件范例 38

To: Reginald Chambers, Electrical Wholesale, Ltd.

From: Howard Zhang

Subject: Switches and **coils**[1]

Dear Mr. Chambers:

Thank you for sending us your catalog, which included pricing and other details. We are interested in doing business with you. However, we are hoping that we can **negotiate**[2] a better price with you. Regarding the electrical switches (model GH-8765), we are **prepared**[3] to **offer**[4] you $1.19 for each switch.

In addition, we are interested in the heating coils that we discussed. In the catalog, they are **listed**[5] at $11.30 per coil. How would you feel about $10.00 per coil? Due to our **finances**[6], the best we can offer is this price. I look forward to your reply concerning this matter.

Regards,
Howard Zhang

中译

收件人：雷金纳德·钱伯斯电器批发有限公司
发件人：霍华德·张
主　题：开关和线圈

亲爱的钱伯斯先生：

感谢您发来贵公司的目录，当中包含了价格及其他详细情况。我们对和贵公司做生意颇感兴趣。然而，我们希望可以和您协商出更好的价格。关于电器开关（型号 GH-8765），本公司愿意出价每组 1.19 美元。

此外，我们也对所讨论的加热线圈感兴趣。在目录中，加热线圈列出的价格是每盘 11.30 美元。您觉得每盘 10 美元如何？基于我们的财务状况，这是我们能提出的最优惠的价格。我很期待您关于此事的回复。

诚挚的祝福

霍华德·张

To: Gerald Livingston

From: May Xu

Subject: Your interest in my company

Dear Mr. Livingston:

It was interesting to talk to you yesterday. During our conversation, you mentioned your interest in making an **investment**[7] in my company. Before we had that discussion, I hadn't given a great deal of thought to the matter of **outside**[8] investors. However, after our meeting, I **took** what you said **into consideration**[9]. In fact, I am now interested in hearing more about your proposal. What did you **have in mind**[10] regarding an investment? Could you tell me what your offer is?

Thank you for your interest in my company, and I am eager to hear more details.

Yours truly,
May Xu

中译

收件人：杰拉德·李文斯顿
发件人：梅·许
主　题：您对本公司的兴趣

亲爱的李文斯顿先生：

昨天与您谈话实在有趣。我们谈话时，您提到了有兴趣投资本公司一事。在我们进行那次讨论之前，我对有关外来投资者的事没有考虑太多。然而，我们会面之后，我考虑了您所说的。事实上，我现在很有兴趣想多听听您对此事的意见。对于这项投资，您心里有什么想法？ 您能告诉我您能提供的报价吗？

感谢您对本公司的兴趣，我很渴望听到更多详细情况。

梅·许　敬上

Vocabulary and Phrases

1. **coil** [kɔɪl] *n.* (金属线、绳索等的)线圈 & *vt.* 使缠绕
   例: The little girl's hair is coiled on top of her head.
   (那个小女孩的头发盘在头顶上。)

2. **negotiate** [nɪˈgoʃɪˌet] *vt.* & *vi.* 商定；达成协议
   negotiate with... 与……协商
   例: The detective managed to negotiate the release of the hostages.
   (那位刑警设法达成了释放人质的协议。)
   The government refused to negotiate with terrorists.
   (政府拒绝和恐怖分子协商。)

3. **prepared** [prɪˈpɛrd] *a.* 愿意的
   be prepared to V 愿意……
   = be willing to V
   例: How much are you prepared to pay?
   (你愿意出多少钱？)

4. **offer** [ˈɔfɚ] *n.* 报价
   例: They made me an offer I couldn't refuse.
   (他们提出了一个使我不能拒绝的报价。)
   I decided to accept their original offer, but it had been withdrawn.
   (我决定接受他们最初的报价，但那项报价已被撤销了。)

5. **list** [lɪst] *vt.* 列入价目表
   例: This tablet computer is listed at $800.
   (这台平板电脑在价格表上是 800 美元。)

6. **finances** [ˈfaɪnænsɪz] *n.* (个人、组织、国家的)财力，财源(常用复数)
   例: It's about time we sorted out our finances.
   (是时候整顿我们的财务了。)

7. **investment** [ɪnˈvɛstmənt] *n.* 投资
   make an investment in... 投资……

8. **outside** [ˈautˌsaɪd] *a.* 不属于本团体(或机构、国家等)的；外部的
   例: We came to the conclusion that we would use an outside consulting firm.
   (我们得出结论会使用外面的咨询公司。)

9. **take... into consideration**
   将……进行考虑
   例: I'll take your advice into careful consideration.
   (我会仔细考虑你的建议。)

10. **have...in mind**
    心中有适当人选(或想做的事等)
    例: Do you have an ideal candidate in mind for this job?
    (你心里有没有做这份工作的理想人选？)

Business Writing Exercises

请按括号中的提示将下列句子译成英文。

1. 我们愿意给你提供单价 10 美元的价格。(We are prepared to...)
   ____
2. 你觉得 15000 美元的报价如何？(How would you feel about...?)
   ____
3. 我们能提供的最优惠价格是 105 美元。(The best price...)
   ____
4. 关于总价，你们有何建议？(What do you propose...?)
   ____
5. 关于我们的大宗订单，你们有何想法？(What do you have in mind...?)
   ____

Unit 2

Setting Conditions & Making a Counter Offer
设定条件与卖方还价

Basic Structure 基本结构

1. 表明感谢买方或卖方愿意出的价格。(We appreciate your offer of...)
2. 要求买方增加订量以获得折扣优惠。(We can only agree to that offer on the condition that you place a sizable order.)
3. 说明付费方式及期限。(We must stipulate that payment be made within 45 days / 30 days of delivery.)

- 这类电子邮件的目的是和对方协商报价的条件。在设定条件时，可以用“On (the) condition that...”（在……条件下）或者“As long as...”（只要……）之类的话来告知对方我方的条件。
- 在相互协商时，能否提出诱因吸引对方是很重要的。可供还价的条件除了有订购的数量之外，延长或缩短付款期限也是其中一环。

Sentence Patterns 写作句型 39

设定条件一

on (the) condition that... 在……条件下

- I can lower the price on the condition that you pay all delivery costs.
  我可以降低价格，但条件是由你支付所有运送费用。

As long as... 只要……

- As long as you guarantee the quality, we can do business together.
  只要你保证质量，我们就能合伙做生意。

设定条件二（较强烈）

We must insist that... / We must insist on sth 我们必须坚持……

- We must insist on an upfront payment of 25 percent.
  我们必须坚持 25% 的预付款。
  * upfront [ˌʌpˈfrʌnt] *a.* 预付的；预交的

We must stipulate that... 我们必须明确要求……

- I must stipulate that your products meet our specifications exactly.
  我必须明确要求你们的产品要完全符合本公司的规格。

设定条件三（较委婉）

If you (do sth), we will (do sth). 假如你……，我们将会……

- If you give us a 15 percent discount, we will increase our order.
  假如你给我们 85 折的优惠，我们将会增加订量。

If we (do sth), will you (do sth)? 假如我们……，你是否会……？

- If we make you our exclusive agent, will you lower your commission?
  假如我们让你成为我们的独家代理商，你是否会降低佣金？
  * exclusive [ɪkˈsklusɪv] *a.* 专有的，独有的
  commission [kəˈmɪʃən] *n.* 佣金；回扣

还价一

How about (price) instead? 改为（某价格）如何？

- How about a total payment of $50,000 instead?
  总金额改为 5 万美元如何？

Would you accept (price)? 你会接受（某价格）吗？

- Would you accept $5.50 per unit?
  你会接受每组 5.5 美元的价格吗？

还价二

I would be more comfortable with (price). 对于（某价格）我会更自在。

- I would be more comfortable with a price of $2.25 for each adapter.
  对于每个电容器 2.25 美元的价格我会更自在。

A price of (amount) seems more reasonable to us. ……的价格对我们而言似乎更合理。

- A price of $18,000 seems more reasonable to us.
  18000 美元的价格对我们而言似乎更合理。

Model E-mail 电子邮件范例 39

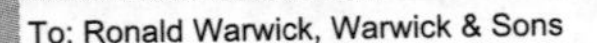

To: Ronald Warwick, Warwick & Sons

From: Peter Deng, OfficeWare International, Inc.

Subject: Stationery **kits**[1]

Dear Mr. Warwick:

We appreciate your offer of $2.79 for each of the stationery kits. Your offer would require us to provide you with a 20 percent discount, and we can only agree to that amount on the **condition**[2] that you **place a sizable**[4] **order**[3] with us. **As long as**[5] your **minimum**[6] order is 10,000 items, we can agree to the price you suggested. In addition, we must **stipulate**[7] that payment be made within 45 days of delivery.

Best regards,
Peter Deng
OfficeWare International, Inc.

To: Peter Deng, OfficeWare International, Inc.

From: Ronald Warwick, Warwick & Sons

Subject: Re: Stationery kits

Dear Mr. Deng:

Thank you for your **detailed**[8] reply. Regarding your proposal, if we increase our order to 15,000, would you accept 90 days for the payment period? We would be more **comfortable with**[9] those conditions. Please let us know if these terms are acceptable to you.

Sincerely,
Ronald Warwick
Warwick & Sons

中译

收件人：罗纳德·沃里克
　　　　沃里克父子公司
发件人：彼得·邓
　　　　办公用品国际有限公司
主　题：文具套装

亲爱的沃里克先生：

　　我们非常感谢您对每个文具套装愿意出2.79美元的报价。贵公司的报价就是要我们提供8折的优惠，我们可以答应这样的优惠额，但前提是贵公司要向我们下大笔订单。只要贵公司的最低订单量为1万套，我们就可以接受贵公司建议的价格。此外，我们必须明确要求货款在交货后45天之内支付。

　　诚挚的祝福

办公用品国际有限公司
彼得·邓

中译

收件人：彼得·邓
　　　　办公用品国际有限公司
发件人：罗纳德·沃里克
　　　　沃里克父子公司
主　题：回复：文具套装

亲爱的邓先生：

　　感谢您详细的回复。关于您的提议，假如本公司订单量增加到1.5万套，您是否会接受货到后90天的付款期？那些条件会让我们感到轻松些。请让我们知道贵公司是否可以接受这些条件。

沃里克父子公司
罗纳德·沃里克　敬上

Vocabulary and Phrases

1. **kit** [kɪt] *n.* 成套工具；成套设备
   a tool kit　一套工具
   a first-aid ki　一套急救用品
2. **condition** [kənˋdɪʃən] *n.* 前提
   on (the) condition that...
   在……的前提下
   例: I'll come to the party on condition that I don't have to wear that chicken costume.
   （我会参加那场派对，前提是不必穿那套鸡的服装。）
   * costume [ˋkɑstjum] *n.*（戏剧或电影的）戏装
3. **place an order**　下订单
   例: I would like to place an order for ten copies of this book.
   （这本书我想订购 10 本。）
4. **sizable** [ˋsaɪzəbl̩] *a.* 相当大的
5. **as long as...**　只要……
   例: We'll go for a walk as long as the weather is good.
   （只要天气好我们就去散步。）
6. **minimum** [ˋmɪnəməm] *a.* 最小量的 & *n.* 最小量
   例: The work was done with the minimum amount of effort.
   （做这项工作没费什么劲。）
   You need to practice playing the piano for <u>a minimum of</u> two hours a day.
   （你需要每天最少练习弹钢琴两小时。）
7. **stipulate** [ˋstɪpjəˏlet] *vt.* 规定；明确要求
   例: A delivery date is stipulated in the contract.
   （这份合约中规定了交货日期。）
8. **detailed** [ˋditeld] *a.* 细节的
   a detailed description / analysis / study
   详细的描述/分析/研究
9. **be comfortable with...**
   对……感到轻松的
   例: The computer geek is more comfortable with computers than with people.
   （比起与人相处，那个电脑鬼才与电脑打交道更能应付自如。）
   * geek [gik] *n.* 对某种事物有深入研究的怪才

Business Writing Exercises

请按括号中的提示将下列句子译成英文。

1. 我们可以降低价格，条件是你们要付现金。(...on condition that...)
   ____
2. 我们坚持要求 10% 的预付款。(an upfront payment)
   ____
3. 如果你们给我们 8 折优惠，我们将增加订货数量。
   ____
4. 你们能接受单价 (the unit price) 为 12.6 美元的价格吗？
   ____
5. 请让我们知道你们是否可以接受这些条件 (these terms)。
   ____

Unit 3

Accepting and Rejecting Offers
接受与拒绝报价

Basic Structure 基本结构

1. 经过仔细考量后，接受对方的报价。(After careful consideration, we are willing to accept your offer.)
2. 针对对方需要更多折扣的问题，表示需要更多时间考虑。(Regarding the discount you are asking for, I need more time to consider your request.)
3. 坚决拒绝对方不合理的报价。(With regard to...., I'm afraid your offer is unacceptable.)

- 要回信拒绝对方的报价时，在信中应保持礼貌，在解释拒绝理由的同时可提出还价（counter-offer，指任一方对报价内容或条件提出修改或要求增加若干条件）。
- 如要表示对报价需要多点时间考虑，可以用“We / I need more time to consider your offer.”（我们 / 我需要更多时间来考虑你的报价。）或者“We'll / I'll have to get back to you on this soon.”（我们 / 我会很快再给你对于此事的答复。）之类的话来告知对方。

Sentence Patterns 写作句型 40

接受对方提出的报价（一）

We're / I'm willing to accept your offer on... 我们/我很乐意接受你对……的报价。

- We're willing to accept your offer on the components.
  我们很乐意接受你对零件的报价。

It sounds like we've got a deal on... 这听起来像是我们对……已经达成协议了。

- It sounds like we've got a deal on the computer accessories.
  这听起来像是我们已在计算机配件方面达成协议了。

接受对方提出的报价（二）

We / I can go along with... 我们/我赞同……

- I can go along with what you are offering.
  我赞同你提出的报价。
  * go along with... 赞同……

We / I believe your offer of (amount) is acceptable. 我们/我认为你（某金额）的报价可以接受。

- We believe that your offer of $10,000 is acceptable.
  我们认为你提出的 1 万美元的报价可以接受。

延迟 / 拖延

We / I need more time to consider... 我们/我需要更多时间来考虑……

- We need more time to consider your offer, I'll give you an answer soon.
  我们需要更多时间来考虑你的报价，我很快会给你答复。

We'll / I'll have to get back to you on... 我们/我得再给你关于……的答复。

- I'll have to get back to you on that. Is next week OK?
  我会再给你关于那件事的答复。下星期可以吗？

拒绝对方提出的报价（一）

We're / I'm afraid (amount) is unacceptable. 恐怕（某金额）不是我们/我所能接受的价格。

- I'm afraid $7 per unit is unacceptable to us.
  恐怕每件 7 美元不是我们能接受的价格。

Unfortunately, your offer is not good enough. 很遗憾，贵公司的报价不够优惠。

- Unfortunately, your offer is not good enough to be considered seriously.
  很遗憾，贵公司的报价不够优惠，我方无法认真考虑。

拒绝对方提出的报价（二）（更坚决）

Your offer of (amount) is out of the question. 你（某金额）的报价是不可能的。

- Your offer of $250,000 is out of the question.
  你 25 万美元的报价是不可能的。
  * be out of the question = be impossible 不可能的

I'm sorry, but your offer isn't even in the ballpark. 我很抱歉，但是你的报价太离谱了。

- I'm sorry, but your offer of $1.87 per screw set isn't even in the ballpark.
  很抱歉，但是针对每枚螺丝钉 1.87 美元的报价，我们觉得太离谱了。

* ballpark 字面上的意思指“棒球场”，商场上则用来指“（数额的）可变通范围”。in the ballpark 则表“在合理范围内”。

Model E-mail 电子邮件范例 40

To: Tiffany Fang

From: Arlene Manz, Home Supplies (Taiwan), Inc.

Subject: Wood products

Dear Ms. Fang:

After considering your proposal, I'm happy to say that we are willing to accept your offer on the wood paneling. I can **go along with**[1] a price of $15 per **tile**[2]. So, it sounds like we have a deal on that.

Regarding the **picture frames**[3], however, I need more time to consider your offer. I need to talk to our production manager more about it because you are asking for a **substantial**[4] discount. I'll have to **get back to**[5] you on that with an answer tomorrow.

Finally, **with regard to**[6] the kitchen cabinets, I'm afraid your offer of $2,500 is unacceptable. At that price, it is just not **profitable**[7] for us at all. Perhaps if you come back with **a** higher **counter offer**[8], we can reach **agreement**[9]. I look forward to discussing this matter more with you.

Sincerely,
Arlene Manz
Home Supplies (Taiwan), Inc.

中译

收件人：蒂芙尼・方
发件人：阿琳・曼兹　家居用品（台湾）有限公司
主　题：木制品

亲爱的方女士：

考虑过你的方案后，我很高兴在此表示本公司乐于接受你对木镶板的报价。我同意你每片 15 美元的价格。所以，这听起来好像我们已对此达成协议。

然而，关于画框的部分，我需要更多时间来考虑你的报价。因为你要求的折扣相当大，所以我需要和生产部经理再讨论一下此事。我得明天才能再回复你此事。

最后，有关厨房橱柜一事，恐怕 2500 美元不是我能接受的价格。以那种价位，本公司就一点利润也没有了。也许如果你出价再高一点，我们就能达成共识了。我期待与你再讨论一下此事。

家居用品（台湾）有限公司
阿琳・曼兹　敬上

Vocabulary and Phrases

1. **go along with...**
   赞同某事；和某人观点一致
   例: I don't go along with Tom's views on abortion.
   （在堕胎问题上我不同意汤姆的观点。）
2. **tile** [taɪl] *n.*瓷砖，瓦片，片状材料（贴墙或铺地用的）
   floor tiles　地砖
3. **picture frame**　相框，画框
4. **substantial** [səbˈstænʃəl] *a.* 大量的 (= considerable [kənˈsɪdərəbl̩])
   例: The first draft of your article required a substantial amount of rewriting.
   （你这篇文章的初稿需要大量的改写。）
5. **get back to sb**　回复某人
   例: I'll find out and get back to you as soon as possible.
   （我会找出答案并尽快回复你。）
6. **with regard to...**　有关……
   例: I'm writing to you with regard to your e-mail of May 1.
   （我写信给你是有关你5月1日发给我的那封电子邮件。）
7. **profitable** [ˈprɑfɪtəbl̩] *a.* 有利润的
   * profit [ˈprɑfɪt] *n.* 利润
   例: Over the past five years, the company has developed into a highly profitable business.
   （过去这5年来，那家公司已发展成为了一家高盈利的企业。）
8. **a counter offer**　还价
   例: The customer rejected the initial offer and made a counter offer.
   （这个客户拒绝了一开始的报价并做出还价。）
9. **agreement** [əˈgrimənt] *n.*（意见或看法）一致（不可数）；协议（可数）
   reach agreement　意见一致
   reach an agreement　达成协议
   例: To our disappointment, the two sides failed to reach agreement.
   （令我们失望的是，双方未能取得一致意见。）
   An agreement was finally reached between management and the union.
   （资方和工会双方终于达成了协议。）

Business Writing Exercises

请按括号中的提示将下列句子译成英文。

1. 我们很乐意接受你们的条件。（accept your terms）
   ____
2. 我想我们可以赞同你们提出的报价。（I think we can go along with...）
   ____
3. 关于价格问题，我会尽快给你回复。（get back to you）
   ____
4. 我们需要多一点时间来考虑你们的报价。
   ____
5. 你们的估价太离谱了。（Your estimate is not even...）
   ____

Unit 1

Budgeting 预算

Basic Structure 基本结构

1. 说明预算案包含内容。
2. 明确说明收入及支出数字。(Our revenues are forecasted to amount to...)(Our expenditures are projected to be...)
3. 总结说明年度赤字或盈余。(The estimated surplus is expected to be...)(Our deficit is expected to reach...)

- 这类电子邮件的目的是向对方概述预算情况，因此内容可包括：
  1. 收入和支出 (revenues and expenditures)。
  2. 盈余或赤字 (surplus or deficit)。
- 在邮件开头可以用 "This draft budget shows the breakdown of..." (本预算草案显示……的明细。) 和 "Here is a preliminary budget showing the allotment of..." (以下是显示……分配的初步预算案。) 之类的话来向对方概述预算案的内容。

Sentence Patterns 写作句型 41

This draft budget shows the breakdown of...
本预算草案显示……的明细。

- This draft budget shows the breakdown of expenditures for the first quarter.
  本预算草案显示第一季度支出的明细。
  * draft [dræft] *n.* 草稿；草案
  breakdown [ˈbrekˌdaʊn] *n.* 数字细目

Here is a preliminary budget showing the allotment of...
以下是显示……分配的初步预算案。

- Here is a preliminary budget showing the allotment of financial resources.
  以下是一个显示财务资源分配的初步预算案。
  * preliminary [prɪˈlɪməˌnɛrɪ] *a.* 初步的
  allotment [əˈlɑtmənt] *n.* 分配

Revenues are forecasted to amount to (amount). 营收预计可达（某金额）。

- Revenues are forecasted to amount to $10 million.
  营收预计可达 1000 万美元。
  * amount to... 总计；共计……

Our income is estimated at (amount) for this time frame.
这段时间内，我们的收入预估为（某金额）。

- Our income is estimated at $6.7 million for this time frame.
  这段时间内，我们的收入预估为 670 万美元。
  * time frame （用于某事的）一段时间

Our expenditures are projected to be (amount). 我们的支出预计为（某金额）。

- Our expenditures are projected to be $1 million.
  我们的支出预计为 100 万美元。

The costs for this period are expected to be (amount).
这段时间的费用预计为（某金额）。

- The costs for this period are expected to be roughly $450,000.
  这段时间的费用预计大约是 45 万美元。

We / I predict there will be a shortfall of (amount). 我们/我预测会有（某金额）的赤字。

- We predict there will be a shortfall of $165,000.
  我们预测会有 16.5 万美元的赤字。
  * shortfall [ˈʃɔrtˌfɔl] *n.* 亏空，赤字（= deficit [ˈdɛfəsɪt]）

The estimated deficit is expected to reach (amount). 估计赤字将达到（某金额）。

- The estimated deficit is expected to reach half a million dollars.
  估计赤字将达到 50 万美元。

Our surplus is expected to be (amount). 我们的盈余预计将会是(某金额)。

- Our surplus is expected to be more than $200,000.
  我们的盈余预计将超过 20 万美元。

We expect revenues to exceed expenditures by (amount).
我们预计收入会超过支出达(某金额)。

- We expect revenues to exceed expenditures by at least $825,000.
  我们预计收入会超过支出至少达 82.5 万美元。
  * by 在此表示“以……的差距”之意。

Model E-mail 电子邮件范例 41

To: Terrance Jones, General Manager

From: Franklin Luo, Financial Controller

Subject: **Preliminary**[1] budget

Dear Mr. Jones:

Here is a preliminary budget for Q1 of next year. This **draft**[2] budget shows the **breakdown**[3] of the **various**[4] **expenditures**[5] and revenues for that period. As you will see in the budget document, revenues are **forecasted**[6] to **amount to**[7] $3,510,000. **In contrast**[8], the costs for this period are expected to be $1,067,000. Therefore, we expect revenues to **exceed**[9] expenditures by $2,443,000.

Best regards,
Franklin Luo

中译

收件人：特伦斯·琼斯　总经理
发件人：富兰克林·罗　财务主管
主　题：初步预算

亲爱的琼斯先生：

以下是明年第一季度的初步预算。本预算草案显示那段时间的各项支出及收入明细。正如您将在预算文件中所见，预计收入达 351 万美元。相比之下，这段时间的预计支出为 106.7 万美元。因此，我们预计收入会超过支出 244.3 万美元。

诚挚的祝福

富兰克林·罗

To: Jack Guo, CEO

From: Kathy Linden, Finance Director

Subject: Draft budget

Dear Mr. Guo:

Please find a copy of the proposed budget for the second half of this year attached to this e-mail. Allow me to **highlight**[10] some of the budget details in this e-mail for your convenience. Our income for the final six months of this **fiscal year**[11] is **estimated**[12] at $3.1 million. However, our expenditures are **projected**[13] to be $3.75 million. As a result, I predict there will be a **shortfall**[14] of $650,000. **In light of**[15] this, you may want to look at some areas where expenses could be reduced.

Yours truly,
Kathy Linden

中译

收件人：杰克·郭　首席执行官
发件人：凯西·林登　财务主管
主　题：预算草案

亲爱的郭先生：

我在这封电子邮件内附上今年下半年拟议预算案的附件。为了您的方便，请容我在这封电子邮件中强调预算案的一些细节。本公司会计年度最后 6 个月的收入预估为 310 万美元。然而，我们的支出预计为 375 万美元。因此，我预测会有 65 万美元的亏损。有鉴于此，您可能会想看一下我们在哪些地方可以削减开支。

凯西·林登　敬上

Vocabulary and Phrases

1. **preliminary** [prɪˈlɪməˌnɛrɪ] *a.* 初步的
   preliminary results / findings
   初步结果／发现
   例: We decided to change the design based on our preliminary findings.
   （基于初步的调查结果，我们决定改变设计。）
2. **draft** [dræft] *n.* 草稿；草案
   the first / final draft　初稿／定稿
   例: I've made a rough draft of my proposal.
   （我已经写好提案的草稿。）
3. **breakdown** [ˈbrekˌdaʊn] *n.* 明细表，分类
   a breakdown of the costs
   成本的详细数字
4. **various** [ˈvɛrɪəs] *a.* 各式各样的
5. **expenditure** [ɪkˈspɛndɪtʃɚ] *n.* 开支，花费（与介词 on 连用）
   例: The government's annual expenditure on education has been reduced.
   （政府在教育上的年度支出被削减了。）

6. **forecast** [ˈfɔrˌkæst] *vt.* 预测（三态同形，或 forecast、forecasted、forecasted）

例: Temperatures were forecast / forecasted to reach 40˚C three days in a row.
（预报温度将连续 3 天达到摄氏 40 度。）

7. **amount to...** 总计……

= add up to...

例: The actor's earnings are said to amount to $40,000 every month.
（据说那位演员的收入每个月总计达 4 万美元。）

8. **in contrast (to / with...)**

（与……）相比之下

例: Our company has expanded enormously, while our competitors, in contrast, have declined.
（我们公司快速扩张，相比之下，我们的竞争对手反而衰退了。）

9. **exceed** [ɪkˈsid] *vt.* 超过

exceed sb's expectations 超过某人的期望（expectations 常用复数）

例: The success of our ad campaign exceeded our expectations.
（这次广告活动的成功超出了我们的预期。）

10. **highlight** [ˈhaɪˌlaɪt] *vt.* 强调

例: The report highlights the need for improved safety regulations for night clubs.
（这份报告强调了对夜店加强安全管理的需求。）

11. **fiscal year** 会计年度

12. **estimate** [ˈɛstəˌmet] *vt.* 估计

例: The event organizer estimated the attendees at 300.
（活动主办者估计参与人数为 300 人。）

The deal is estimated to be worth around $50,000.
（这笔交易估计价值5万美元左右。）

13. **project** [prəˈdʒɛkt] *vt.* 预计

例: The unemployment rate has been projected to fall.
（预计失业率将下降。）

14. **shortfall** [ˈʃɔrtˌfɔl] *n.* 亏损；短缺

a shortfall in food supply
食物供给短缺

15. **in light of...** 有鉴于……

例: In light of recent incidents, customers should take particular care of their personal belongings.
（鉴于最近发生的事件，顾客应该特别小心自身的物品。）

Business Writing Exercises

请按括号中的提示将下列句子译成英文。

1. 营收预计可达 500 万美元。（Revenues are forecasted to...）

____

2. 我们的支出（our expenditures）预计为 200 万美元。

____

3. 预估赤字（the estimated deficit）将达 1000 万美元。

____

4. 我们的盈余（our surplus）预计将超过 100 万美元。

____

5. 我们预计收入会超过支出达 300 万美元。

____

Unit 2

Discussing Profit and Loss
讨论盈利与亏损

Basic Structure 基本结构

1. 概括指出整体盈利与亏损状况。（To summarize, the company earned a profit this year.）
2. 详细指出各部门的盈亏情况。（The division suffered a loss.）（The division broke even this year.）
3. 提供可能改善亏损的方法。

- 这类电子邮件的目的是向股东说明各部门的盈利与亏损状况，因此内容可包括：
  1. 获利及亏损（profit and loss）情况。
  2. 损益平衡（break even）情况。
- 如要说明产品获利情况，可以用“The profit margin is high / low on...”（……的获利率很高 / 低。）或者“The markup on (sth) is high / low / (amount).”（（某物）的加价很高 / 很低 / 是某额度）之类的话来表示。

Sentence Patterns 写作句型 42

盈利

The company made / earned a profit of... 该公司赚取……的利润。

- The company made a profit of $9 million this year.
  该公司今年盈利 900 万美元。

The product is / was profitable for... 那项产品对……而言是可获利的。

- Our popular notebook was profitable for the company.
  本公司畅销的笔记本电脑对公司而言是可获利的产品。

亏损

(Sb) suffered / made a loss... （某人）亏损了……

- We suffered a loss of $1.8 million in the second quarter.
  我们在第二季度亏损了180 万美元。

损益平衡

(Sb) broke even... （某人）……收支平衡。

- The company just about broke even last year.
  这家公司去年只接近收支平衡。

(Sb) made neither a profit nor a loss... （某人）收支平衡……

- The company made neither a profit nor a loss this year.
  这家公司今年收支平衡。

形容高额盈利和亏损的惯用语

(Sb) made a killing... （某人）发大财……

- We made a killing after we launched our new product.
  新产品投放市场后，我们发了大财。
  * launch [lɔŋtʃ] *vt.* （产品）投放市场

(Sb) lost (one's) shirt... （某人）赔得精光……（字面意思为：某人赔钱赔到连衬衫都赔掉了）

- The automobile company nearly lost its shirt on the unpopular model.
  这家汽车公司在这款不畅销的车型上几乎赔得精光。

商品利润

The profit margin is high / low on... ……的获利率很高/低。

- The profit margin is low on this product, so we're not interested in selling it.
  该产品的获利率很低，所以我们没有兴趣销售它。

The markup on (sth) is high / low / (amount). （某物）的加价很高/很低/是某额度。

- The store's markup on the camera is $150.
  这家店对那台相机加价了 150 美元。
  * markup [ˈmɑrkˌʌp] *n.* 加价（成本价与销售价之间的差额）

Model E-mail 电子邮件范例 42

To: **Shareholders**[1] of Nature's Best Foods, Ltd.

From: The board of directors

Subject: Annual report

Dear Shareholders:

As the CEO of Nature's Best Foods, Ltd, it's my pleasure to **present**[2] you with our annual report, which is attached to this e-mail. To **summarize**[3], the company earned a **profit**[4] in almost all of its **divisions**[5] this year. In particular, our health food products were quite profitable for us. However, the **organic**[6] snack food division **suffered**[7] a loss. In addition, our breakfast cereal division only **broke even**[8] this year.

Despite the poor performance from organic snacks and breakfast cereals, we are quite **optimistic**[9] about several of our new products that will be **launched**[10] soon. We may not exactly **make a killing**[11] on them, but we're confident they will be extremely profitable. You will be glad to know that the profit margins on these new items are high.

The entire board and I all wish you a happy and healthy New Year.

Sincerely,
Kenneth Tu

中译

收件人：大自然最佳食品有限公司所有股东

发件人：董事会

主　题：年度报告

亲爱的股东们：

身为大自然最佳食品有限公司的首席执行官，我很荣幸向诸位呈献本公司的年度报告，该报告附在这封电子邮件内。概括而言，本公司今年几乎所有部门都赚钱了，尤以健康食品对本公司而言获利相当高。但是有机零食部门却蒙受损失。此外，本公司的谷类早餐部门今年仅收支平衡。

尽管有机零食部和谷类早餐部表现不佳，我们还是很看好即将投放市场的几款新产品。我们或许不可能靠新产品就发大财，但我们相信新产品会非常盈利。诸位将很高兴得知，这些新产品的获利率很高。

我和全体董事会成员祝诸位新年快乐、身体健康。

肯尼斯·涂　敬上

Vocabulary and Phrases

1. **shareholder** [ˋʃɛr,holdɚ] *n.* 股东
2. **present** [prɪˋzɛnt] *vt.* 赠送，呈献
   present sb with sth
   将某物赠送/呈献给某人
   例: The brave young man was presented with a medal.
   (这位英勇的年轻人获颁一枚勋章。)
3. **summarize** [ˋsʌmə,raɪz] *vt.* 概述
   例: The findings of the survey are summarized at the end of the chapter.
   (在这章末尾对调查结果做了总结。)
4. **profit** [ˋprɑfɪt] *n.* 利润；收益
   make a huge profit　赚得厚利
   at a profit　以获利的方式
   例: The company made a huge profit of $12 million last year.
   (该公司去年赚得1200万美元的厚利。)
   We should be able to sell the house at a profit.
   (我们卖掉这栋房子应该可以获利。)
5. **division** [dəˋvɪʒən] *n.* (机构的)部门
6. **organic** [ɔrˋgænɪk] *a.* 有机的
7. **suffer** [ˋsʌfɚ] *vt.* 承受 & *vi.* 长期受某疾病所苦(与 from 连用)
   例: The company suffered huge losses due to a bad investment.
   (那家公司由于投资不善而蒙受巨额亏损。)
   Ted suffers from asthma.
   (泰德患有哮喘。)
8. **break even**　损益平衡，打平
9. **optimistic** [,ɑptəˋmɪstɪk] *a.* 乐观的
   be optimistic about...　对……表示乐观
   例: I think you're being a little optimistic about the outcome of the negotiations.
   (我发现你对谈判的最终结果有点乐观。)
10. **launch** [lɔntʃ] *vt.* (首次)上市，发行
    例: The new product is scheduled to be launched next month.
    (新产品预计在下个月上市。)
11. **make a killing**　发大财

Business Writing Exercises

请按括号中的提示将下列句子译成英文。

1. 我们公司平均一年赚取500万美元的利润。(On average,...)
   ____
2. 他们在第三季度亏损了1000万美元。(They suffered a loss...)
   ____
3. 我们公司去年不赚也不赔。(neither made a profit...)
   ____
4. 他们卖有机蔬菜和水果发了大财。(They made a killing...)
   ____
5. 很多人在股票市场赔得很惨。(lose one's shirt)
   ____

Unit 1

Discussing Sales Targets
讨论销售目标

Basic Structure 基本结构

1. 来信说明销售会议的结论。(I want to inform you of the results of the sales meeting.)
2. 讨论上一季度的销售业绩。(We exceeded our sales objectives for last quarter.) (We fell short of our sales targets for last quarter.)
3. 预估下一季度的销售业绩。(The sales forecast for the next quarter shows...)

- 这类电子邮件的目的是让对方大致了解销售会议的内容,因此内容可包括:
  1. 上一季度的销售结果(sales results)。
  2. 新一季度的销量预估(new sales forecast)。
- 如要叙述某段时间内的销售情况,可以用"The sales volume for (period of time) was..."((某段时间)的销售量为……)或者"The sales turnover during (period of time) was..."((某段时间)的销售营业额是……)来表示。

Sentence Patterns 写作句型 43

...meet / exceed our sales objectives. ……达到/超过我们的销售目标。

- With hard work, we were able to exceed our sales objectives.
  由于工作努力，我们才能超过我们的销售目标。
  * sales objectives 销售目标

...reach / surpass our sales goals. ……达到/超越我们的销售目标。

- Let's try hard to surpass our sales goals.
  咱们一起努力来超越我们的销售目标吧。

...fall short of our sales targets. ……未达到我们的销售目标。

- Unfortunately, we fell short of our sales targets.
  不幸地，我们未达到我们的销售目标。
  * fall short (of...) 未达到；不符合（……）

...miss our sales objectives. ……无法达到我们的销售目标。

- I'm afraid we are going to miss our sales objectives for this quarter.
  恐怕这一季度我们无法达到我们的销售目标。

The sales volume for (period of time) was... （某段时间）的销售量为……

- The sales volume for last month was 1,500 units.
  上个月的销售量为 1500 套。

The sales turnover during (period of time) was... （某段时间）的销售营业额是……

- The (sales) turnover during the last quarter was $7.5 million.
  上一季度的（销售）营业额是 750 万美元。
  * turnover [ˋtɝnˏovɚ] *n.* 营业额

The sales forecast for (period of time) ... 对（某时间）的销售预估……

- The sales forecast for the third quarter has been completed.
  第三季度的销售预估已经完成了。（forecast 为名词）

Sales are forecasted to... 销售量预计……

- Sales are forecasted to increase 10 percent from last year.
  比起去年，今年的销售量预计增加 10%。（forecast 为动词）
  * launch [lɔntʃ] *vt.* （产品）投放市场

...set sales quotas. ……设定销售配额。

- We need to set realistic sales quotas.
  我们需要设定能够实现的销售配额。
  * quota [ˋkwotə] *n.* 配额

...establish sales targets. ……设立销售目标。

- At the meeting, we must establish sales targets for next year.
  在会议上，我们必须为明年设立一个销售目标。

Model E-mail 电子邮件范例 43

To: Gary Liang

From: Martin Lee

Subject: Sales meeting

Hi Gary:

How is your vacation going? I want to let you know about the results of the sales meeting today so you can **be well informed**[1] by the time you get back to work.

First, there was a review of last quarter's sales results. Although we **exceeded**[2] our sales **objectives**[3] for last month, we **fell short of**[4] our sales targets for the entire quarter (the first two months of the year were quite poor, in fact). **Sales volumes**[5] for the first three months of the year were 5,400 units, while **turnover**[6] during that period was $780,000.

The review of last quarter was followed by a presentation of the new sales forecast. The sales forecast for the next quarter shows a 12-percent increase from the same period last year. There will be a meeting next week to establish sales quotas for each member of the sales staff.

See you later this week, Gary.

Best regards,
Martin

中译

收件人：加里・梁

发件人：马丁・李

主　题：销售会议

嗨，加里：

你的假期过得如何？我想告诉你今天销售会议的结果，好让你在回到工作岗位时能较快熟悉状况。

首先，我们针对上一季度的销售结果进行了评述。虽然我们上个月的业绩超过了销售目标，但我们却未能达到整季度的销售目标（其实，今年前两个月的业绩相当差）。今年前3个月的销售量是5400套，而那段时间的营业额则是78万美元。

评述完上一季度的业绩之后，接着就是新一季度的销售量预估简报。下一季度的销售量预估显示将比去年同期增加12%。下星期将有一次会议，会上每位业务员都要设定自己的销售配额。

加里，这个星期晚些时候见。

诚挚的祝福

马丁

Vocabulary and Phrases

1. **be well informed**
   见多识广的；消息灵通的
   例: Most people are not very well informed about the disease.
   （大多数人对于这种疾病了解不多。）
2. **exceed** [ɪkˋsid] *vt.* 超过
   例: The success of the ad campaign exceeded our expectations.
   （这次广告活动的成功超出了我们的预期。）
3. **objective** [əbˋdʒɛktɪv] *n.* 目标
   meet / achieve one's objectives
   达到／实现某人的目标
   例: Can the sales force achieve its sales objectives?
   （销售团队能否达到销售目标呢？）
   You must set realistic aims and objectives for yourself.
   （你必须给自己设定切实可行的目的和目标。）
4. **fall short of...** 未达到……
   例: Their service fell far short of our expectations.
   （他们的服务远没有我们预期的那么好。）
5. **sales volume** 销售量
6. **turnover** [ˋtɝnˏovɚ] *n.* 营业额；人事变更率
   例: The company has an annual turnover of $2 million.
   （该公司的年营业额为 200 万美元。）
   The firm has a high turnover of staff.
   （该公司的人员流动率很高。）

Business Writing Exercises

请按括号中的提示将下列句子译成英文。

1. 我确信我们能够达到明年的销售目标。（meet our sales objectives）
   ____
2. 上个月的销售量没有达到我们的预期。（fall short of our expectations）
   ____
3. 第三季度的营业额是 600 万美元。（The sales turnover for...）
   ____
4. 对于下一季度的销售预估已经完成。（The sales forecast for...）
   ____
5. 我们需要给每一位销售代表（every sales representative）设定销售配额（set sales quotas）。
   ____

Unit 2

Talking about Sales Promotions
谈论促销活动

Basic Structure 基本结构

1. 明确指出促销的原因。(We need to create a buzz about our products.)
2. 提出几种促销活动的可行方案。(We will provide a discount of up to 20 percent on...) (We will conduct a reward program.) (We will offer cash rebates.)
3. 说明讨论促销活动的开会时间及地点。

- 这类电子邮件的目的是要员工想出新的促销方法，因此内容重点在于：
  1. 目前促销的标准做法（standard methods）。
  2. 在特殊情况（at some point）下的做法。
- 老式惯用语会让读者有不好的印象，甚至造成困惑。例如我们会使用“Please remember...”（请记得……）而不是用“Please be reminded...”（在此提醒您……），一封好的电子邮件需要的是精确的信息，而不是过多的修饰语。
- 一条很长且不分段的信息会造成阅读上的困惑。记得用空白来区分各段落，这样不但方便阅读，也能让收件人对发件人有个好印象。

Sentence Patterns 写作句型 44

...launch a sales campaign... ……发起促销活动……

- The company will launch a huge sales campaign in the fall.
  该公司将于今年秋天举办一次大型促销活动。

...run a sales initiative... ……进行销售活动……

- How long has it been since we ran a sales initiative for this product last time?
  自从我们上次为这项产品做销售活动以来已过多久了？
  * initiative [ɪˋnɪʃətɪv] *n.* 倡议；主动的行动

...create a buzz about (sth). ……（为某物）制造话题/引起热议。

- We need to create a buzz about the new mobile phone.
  我们需要为这款新手机制造话题。
  * buzz [bʌz] *n.* 兴奋及期待的心情
  create a buzz = create anticipation and excitement about something

...generate word-of-mouth of advertising... ……制造口头广告……

- Do you have any ideas about how to generate word-of-mouth advertising?
  你对制造口头广告有什么想法吗？

...provide a discount of (amount / percentage) on (sth).
……针对（某物）提供（某数量/某百分比）的折扣。

- We've agreed to provide a discount of 15 percent on the notebooks.
  我们已同意给这些笔记本电脑提供 85 折的折扣。

...offer cash rebates on (sth). ……给（某物）提供现金折扣。

- We should try offering cash rebates on our automobiles.
  我们应试着给本公司的汽车提供现金折扣。
  * rebate [ˋribet] *n.* （现金）折扣

...give away free samples of... ……赠送免费样品……

- The company decided to give away free samples of its new chocolate bar.
  公司决定赠送新款巧克力棒的免费样品。

...conduct a reward program... ……进行积分回馈……

- If we conduct a reward program, do you think it will be effective?
  如果我们进行积分回馈，你认为会有效吗？

...undercut the competition... ……削价与竞争对手抢生意……

- The best way for us to increase sales is to undercut the competition.
  对我们来说，增加销售最好的方法就是与对手削价竞争。

...engage in a price war... ……进行价格战……

- Let's try not to engage in a price war unless there is no other choice.
  除非没有其他选择，咱们就设法不要进行价格战吧。

Model E-mail 电子邮件范例 44

To: All sales staff

From: Ted Higgins, sales manager

Subject: Sales campaign

To all sales reps:

I'm writing to remind you of an important meeting next Thursday. During that meeting, we will **discuss**[1] launching a major sales campaign. As you know, our sales are down this year, and we need to really create a **buzz**[2] about our products. I want you all to think of **innovative**[3] ways to **generate**[4] **word-of-mouth**[5] advertising.

Of course, we are considering all of the **standard**[6] methods that can be used to **stimulate**[7] sales. For example, we are considering providing a discount of **up to**[8] 20 percent on some of our models. We will **conduct**[9] a **reward**[10] program, and perhaps even offer cash **rebates**[11] on a few items.

I want to avoid **undercutting**[12] the **competition**[13] because that will hurt our **profit margins**[14]. However, if we need to **engage**[15] in a price war **at some point**[16], we are prepared to do that. See you all next Thursday.

Regards,
Ted

中译

收件人：所有销售人员
发件人：泰德·希金斯
销售经理
主　题：促销活动

致所有销售代表：

我写信是为了提醒诸位下星期四有一次重要会议。会议上，我们会商讨举行一场大型促销活动。正如诸位所知，今年我们的销售量减少了，我们需要让本公司的产品引起热议。我要诸位想出一些新颖的点子好让我们能制造口头广告。

当然，我们正在考虑所有能用来刺激销售的标准做法。例如，我们正考虑对某些型号的产品提供高达8折的折扣。我们会进行积分回馈，甚至也许会对某些产品提供现金折扣。

我想避免与对手削价竞争，因为这会有损于我们的利润率。然而，如果在某种情况下需要进行价格战时，我们也会做好准备。下星期四见。

诚挚的祝福

泰德

Vocabulary and Phrases

1. **discuss + V-ing** 讨论……
   例: We briefly discussed switching to another supplier at the meeting.
   （我们在会议上简短讨论要换另一家供应商的事。）
2. **buzz** [bʌz] *n.*兴奋及期待的心情
   例: I love cycling fast — it gives me a real buzz.
   （我喜欢骑自行车骑得很快——那让我很兴奋。）
   I really get a buzz out of my work.
   （我真的从工作中得到很大的乐趣。）
3. **innovative** [ˋɪnəˌvetɪv] *a.* 创新的
4. **generate** [ˋdʒɛnəˌret] *vt.* 产生
   例: These measures will increase the company's ability to generate revenues.
   （这些措施会提高该公司创收的能力。）
5. **word-of-mouth** 口头的
   例: The news spread by word-of-mouth.
   （这个消息是口头传开的。）

6. **standard** [ˈstændəd] *a.* 标准的

7. **stimulate** [ˈstɪmjəˌlet] *vt.* 促进

例: The exhibition stimulated the public's interest in the artist's work.
（这个展览刺激了大众对那位艺术家的作品的兴致。）

8. **up to...** 到达（某数量、程度等）

例: The temperature went up to 33 ℃.
（气温上升到了摄氏 33 度。）

9. **conduct** [kənˈdʌkt] *vt.* 执行

conduct an experiment / an inquiry / a survey 进行实验 / 询问 / 调查

10. **reward** [rɪˈwɔrd] *n.* 奖励；回报

a reward program 积分回馈

例: You deserve a reward for being so helpful.
（你帮了这么大的忙，理应受到奖励。）

11. **rebate** [ˈribet] *n.* 现金折扣

= cash rebate

12. **undercut** [ˌʌndəˈkʌt] *vt.* 削价与……抢生意（三态同形）

例: We're planning to undercut our European rivals by 10%.
（我们计划要削价10%与我们的欧洲对手竞争。）

13. **competition** [ˌkɑːmpəˈtɪʃn] 竞争对手

14. **profit margin** 利润率

15. **engage** [ɪnˈgedʒ] *vi.* & *vt.* 从事

engage (sb) in...
（使某人）从事 / 参加……

例: Even in prison, Henry continued to engage in criminal activities.
（亨利甚至在监狱里还继续从事犯罪活动。）

Once Brenda engages you in conversation, you're stuck with her for at least half an hour.
（布伦达一找你聊天，你至少半小时内无法脱身。）

16. **at some point** 在某个时候（point 在此表示"时刻；阶段"）

例: Many people feel lonely at some point in their lives.
（许多人在人生的某个阶段都会感到孤单。）

At this point, I don't care what you decide to do.
（此刻，我不在乎你决定做什么。）

Business Writing Exercises

请按括号中的提示将下列句子译成英文。

1. 我们将在 11 月份进行一次促销活动。（launch a sales campaign）

___

2. 我们需要为我们的新产品制造话题。（create a buzz）

___

3. 他们正针对电器用品（electric appliances）提供现金折扣。（cash rebates）

___

4. 他们正在赠送免费的化妆品样品。（give away free samples）

___

5. 那两家百货商场正在打价格战。（engage in a price war）

___

Unit 1

Writing about Markets and Market Share
撰写市场及市场占有率相关情况

Basic Structure　基本结构

1. 表示已做过市场分析。(I've finished the market analysis.)
2. 确定目标市场以取得竞争优势。(We need to further define our target market.)
3. 说明目前市场供需状况。(The items are still in demand.)

- 这类电子邮件的目的是向对方报告市场分析的结果，因此内容可包括：
  1. 说明目标市场（target market）。
  2. 分析市场需求（market demand）。
  3. 分析市场占有率（market share）。
- 如要说明某市场已经饱和，可以用“The market is saturated with (sth).”（(某物)的市场已经饱和了。）或者“(Sth) has flooded the market.”（(某物)已充斥市场。）来表示。

Sentence Patterns 写作句型 45

...identify / define (one's) target market. ……找出/确定(某人的)目标市场。

- The company took great pains to identify its target market.
  那家公司努力要找出目标市场。
  * take great pains (to V) 费力地(做……)
  = go to great pains (to V)

...reach (one's) target audience. ……引起(某人的)目标客户群的注意。

- What is the best media to use to reach our target audience?
  什么媒体最适合用来引起我们目标客户群的注意?
  * target audience 目标客户群;目标观众

...gain / lose market share. ……取得/失去市场占有率。

- Over the past two years, our company has gained a lot of market share.
  过去两年中,本公司已经取得许多市场占有率。

...stimulate / hurt market growth. ……刺激/损害市场成长。

- I hope the launch of the new product will stimulate market growth.
  我希望新产品上市将刺激市场增长。

a / an (adj.) segment of the market 市场的(某)部分

- Women aged 18 to 35 make up a large segment of the market.
  18 到 35 岁的女性构成了市场的一大部分。
  * make up... 构成……
  segment [ˋsɛgmənt] *n.* 部分

...the market is saturated with (sth). ……(某事物)的市场已经饱和了。

- Analysts say the market is saturated with mini-notebooks.
  分析师指出迷你笔记本电脑的市场已经饱和了。
  * saturate [ˋsætʃəˏret] *vt.* 使饱和

...(sth) has flooded the market. ……(某事物)已充斥市场。

- Low-priced footwear has flooded the market recently.
  低价位的鞋最近已充斥市场。
  * footwear [ˋfʊtˏwɛr] *n.* (总称)鞋类(集合名词,不可数)
  flood [flʌd] *vt.* (使)充斥

...(sth) is in demand. ……(某事物)有需求量。

- Good, innovative products are always in great demand.
  质量优良且有创新的产品需求量总是很大。
  * in demand 有需求量

Model E-mail 电子邮件范例 45

To: Carmen Heinz

From: Albert Han

Subject: Marketing advice

Dear Carmen:

I'm writing to let you know I've finished the **market analysis**[1] that you asked me to complete. You will find the full report attached to this e-mail. However, I'd just like to **summarize**[2] a few points.

Regarding your target audience, I think you were a bit unclear as to who your products are **primarily**[3] **aimed**[4] at. As a result, I've helped you to further define your target market. Two **segments**[5] of the market you **overlooked**[6] were retired men and retired women. Considering the competition in this market, you need to do everything you can to gain **market share**[7].

You asked me if the market is already **saturated**[8] with similar products. Actually, I found that items like yours are still **in demand**[9]. Demand still **outweighs**[10] the supply in this area.

Please let me know when you would like to meet to discuss the results further.

Yours truly,
Albert

中译

收件人：卡门·海因兹
发件人：阿尔伯特·韩
主　题：营销建议

亲爱的卡门：

我来信是为了要告诉你我已经完成了你要求我做的市场分析。你会在这封电子邮件附件中找到完整的报告。不过，我想先为你总结几个要点。

关于贵公司的目标客户群，我认为你们还有点不太清楚你们的产品主要针对谁而设计。因此，我已经帮你们进一步确定了目标市场。贵公司忽略的两部分市场分别是退休男性及退休女性。考虑到这个市场的竞争性，你们要尽一切可能来取得市场占有率。

你问我同类产品的市场是否已经饱和。事实上，我发现像贵公司这样的产品仍然有市场需求量。在这个地区仍是供不应求。

请告诉我你想何时会面以进一步讨论这个结果。

阿尔伯特　敬上

Vocabulary and Phrases

1. **market analysis** 市场分析
   * analysis [əˋnæləsɪs] *n.* 分析
2. **summarize** [ˋsʌməˌraɪz] *vt.* 总结；概括
   例: I'm going to summarize the main points of the argument in a few words.
   （我将用几句话来总结这个论点的要点。）
3. **primarily** [praɪˋmɛrəlɪ] *adv.* 主要地
   (= mainly)
4. **aim** [em] *vt.* 目的是，旨在
   be aimed at... 旨在……
   例: These measures are aimed at preventing violent crime from getting worse.
   （这些措施旨在防止暴力犯罪更加恶化。）
5. **segment** [ˋsɛgmənt] *n.* 部分，片，段
   例: People over the age of 85 make up the fastest-growing population segment.
   （85 岁以上的人构成了人口增长最快的部分。）
6. **overlook** [ˌovɚˋlʊk] *vt.* 忽略；眺望
   例: You seem to have overlooked one important factor.
   （你似乎忽略了一个重要的因素。）
   Our hotel room overlooked the ocean.
   （我们的饭店房间远眺大海。）
7. **market share** 市场占有率
8. **saturate** [ˋsætʃəˌret] *vt.* 使饱和
   例: The company claimed to have saturated the market for smart cellphone.
   （该公司声称他们的智能手机已使市场饱和。）
9. **in demand** 有需求量
   例: Good teachers are always in (great) demand.
   （优秀的老师总是哪里都需要。）
10. **outweigh** [ˌaʊtˋwe] *vt.* 大于，超过
    例: The advantages far outweigh the disadvantages.
    （利远大于弊。）

Business Writing Exercises

请按括号中的提示将下列句子译成英文。

1. 首先（First of all），我们要能够确定我们的目标市场。（define our target market）
   ____
2. 研发工作（Research and development）将帮助我们取得更多的市场占有率。
   ____
3. 青少年消费者占了市场的一大部分。（make up）
   ____
4. 最近，假货（fake goods）已经充斥整个市场。（flood the market）
   ____
5. 自从去年以来，智能手机（smartphones）的需求量一直很大。（in great demand）
   ____

Unit 2

Writing about Brands
撰写品牌打造相关情况

Basic Structure 基本结构

1. 说明打造品牌的重要性。(品牌认同若未建立，可能会导致销售量下滑)(We need to work harder to develop our brand identity and awareness.)
2. 提供建立品牌知名度的各种方式。(We can launch an ad campaign to not only promote our products but also build our brand awareness.)
3. 佐证说明如果未能打造自己的品牌会有何结果。(If we don't build our brand, we will see increased competition from generic brands with lower prices.)

- 这类电子邮件的目的是向对方说明打造品牌的原因及做法，因此内容可包括：
  1. 品牌（brand）的重要性。
  2. 必要的做法（methods）。
- 如要向对方提出建议，可以用“I'd strongly / highly advise + V-ing”(我会强烈建议……）或者“I think I'd advise sb to V”（我想我会建议某人……）来表示。

Sentence Patterns 写作句型 46

a name-brand product 名牌产品
a brand-name product 有品牌的产品
- Do you always buy well-known, name-brand products?
  你总是会买知名的名牌产品吗?

a global brand 全球性的品牌
- Coca-Cola, Apple, and Acer are all global brands.
  可口可乐、苹果和宏碁都是全球性的品牌。

develop brand loyalty 拓展(消费者的)品牌忠诚度
- The company has spent a lot of money on developing brand loyalty.
  这家公司已经花了很多钱拓展(消费者)对该品牌的忠诚度。

build the reputation of a brand 建立品牌声誉
- There are many ways to build the reputation of a brand.
  建立一个品牌的声誉有很多种方式。

position (a product) 定位(某产品)
- We need to better position our fruit drinks on the market.
  我们需要为本公司的果汁饮料找到更好的市场定位。

differentiate A from B 分辨 A 与 B 的差别
- How can we differentiate our snacks from the competition's?
  我们如何将本公司的点心产品和竞争对手的区别开来?
  * differentiate [ˌdɪfəˈrɛnʃɪˌet] *vt. & vi.* 区分, 辨别

I'd strongly / highly advise + V-ing 我会强烈建议……
- I'd strongly advise launching an ad campaign to not only promote our products but also build our brand awareness.
  我强力建议举办一次广告活动, 目的不仅要促销产品, 还要建立品牌知名度。

I think I'd advise sb to V 我想我会建议某人……
- I think I'd advise the manager to use a celebrity to endorse our products.
  我想我会建议经理请名人来为本公司产品代言。
  * endorse [ɪnˈdɔrs] *vt.* (为某产品)代言

a generic product / brand 杂牌产品 / 杂牌
- Generic products are usually cheaper than brand-name products.
  杂牌产品通常比名牌产品要便宜。
  * generic [dʒəˈnɛrɪk] *a.* (产品, 尤指药物)无商标的; 杂牌的

a no-name product / brand 无名产品 / 品牌
- No-name brands often have unattractive packaging.
  无名品牌往往包装不吸引人。
  * unattractive [ˌʌnəˈtræktɪv] *a.* 不漂亮的, 难看的

Model E-mail 电子邮件范例 46

To: Patrick Ma, CEO

From: Wanda Quinlan, Marketing Director

Subject: Branding

Dear Patrick:

You asked me to offer some suggestions to help our **slumping**[1] sales. I've thought about it quite hard and have **come up with**[2] the following advice.

I think the main problem is with our **branding**[3]. Yes, we do offer **brand-name**[4] products, but we are not really a global brand. We need to work harder to develop our brand identity and **awareness**[5]. Once we do that, then I think brand loyalty will naturally follow.

In addition, the company must **position**[6] its products better. We can do that by strongly **differentiating**[7] our products from those of the competition through advertising. For example, we can launch an ad campaign to not only promote our products but also build our brand awareness. Besides, we can use a celebrity to endorse our products. **To this end**[8], I highly advise hiring someone to **take full responsibility**[9] for brand management. In my opinion, a good brand manager will really help us. If we don't build our brand, we will see increased competition from **generic**[10] brands with lower prices.

Yours truly,
Wanda

中译

收件人：帕特里克·马　首席执行官

发件人：旺达·昆兰　营销总监

主　题：品牌打造

亲爱的帕特里克：

您要求我提供一些建议来帮助我们疲软的销售状况。我绞尽脑汁想出了以下建议：

我认为主要的问题在于本公司的品牌打造。没错，本公司确实提供了品牌产品，但我们还不算真正的全球性的品牌。我们需要更努力来拓展品牌识别度和知名度。一旦我们这么做了，那么我认为品牌忠诚度自然会随之而来。

此外，本公司必需在产品定位上做得更完善。我们可以通过广告来强烈区别本公司产品和竞争对手产品的差别，借以达到该目的。举例来说，我们可以进行广告宣传活动，不仅是为了促销产品，还要建立品牌知名度。此外，我们可以请名人来为产品代言。为此，我强烈建议您雇用一位能为品牌管理负起全责的人。依我之见，优秀的品牌经理将对本公司有所帮助。假如我们不打造自己的品牌，我们将看到来自低价位杂牌的竞争日益增加。

旺达　敬上

Vocabulary and Phrases

1. **slump** [slʌmp] *vi. & n.* （价格、价值、数量等）骤降，猛跌

   a slump in sales / profits
   销售量／利润的锐减

   例: The paper's circulation has been slumping over the past five years.
   （这份报纸的发行量过去 5 年来一直在下降。）

   * circulation [ˌsɜːkjəˈleʃən] *n.* （报刊）发行量，销售量

2. **come up with...** 想出（点子、方法）

   例: Ted came up with a good idea to increase sales.
   （泰德想出一个增加销售量的好点子。）

3. **branding** [ˈbrændɪŋ] *n.* 品牌打造（不可数）

4. **brand-name** *a* 商标的；品牌的

   brand name 商标；品牌
   brand identity 品牌识别度
   brand loyalty 品牌忠诚度

5. **awareness** [əˈwɛrnəs] *n.* 意识；了解

   raise / increase public awareness of... 提高公众对……的认识

   例: It's important to raise public awareness of the spread of the disease.
   （提高公众对那种疾病传播的认识是很重要的。）

6. **position** [pəˈzɪʃən] *vt.* 确定……的位置

   例: Our first priority is to position our products in the market.
   （我们的首要任务就是确定产品的市场定位。）

7. **differentiate** [ˌdɪfəˈrɛnʃɪˌet] *vt. & vi.* （使）区别

   differentiate between A and B
   = differentiate A from B 区别 A 与 B

   例: It's difficult for me to differentiate between the twins.
   （对我来说要区分那对双胞胎很困难。）

   The waterproof roof differentiates this house from others in the area.
   （防水屋顶使那栋房子和该地区的其他房子有所区别。）

8. **To this end,...** 为此目的，……

   例: The professor wanted science students to take an interest in the arts. To this end, he offered a course in modern arts.
   （那位教授想要让科学生对艺术产生兴趣。为此目的，他开了一门现代艺术的课。）

9. **take full / overall responsibility (for...)** （对……）负全责

   例: You should take full responsibility for the mistake instead of blaming it on others.
   （你应该对该错误负全责，而不是把责任怪在别人头上。）

10. **generic** [dʒəˈnɛrɪk] *a.* 无商标的，杂牌的

    generic brands 杂牌产品

Business Writing Exercises

请按括号中的提示将下列句子译成英文。

1. 经过多年的努力，我们已成为一家全球性的品牌。

   ____

2. 他们想尽快在客户中建立品牌忠诚度。（establish brand loyalty）

   ____

3. 你能区分真品和假货吗？（differentiate A from B）

   ____

4. 我强烈推荐使用微博来促销（promote sales）。（recommend + V-ing）

   ____

5. 有品牌的产品比杂牌产品更耐用。（more durable than）

   ____

Unit 1

Checking Details 查核细节

Basic Structure 基本结构

1. 提及先前的会面或对话，说明来意。(Regarding our previous discussion, there is something I am unclear about.)
2. 要求再一次解释或澄清之前不清楚之处。(I was wondering if you could explain that concept to me again.)
3. 重述自己对该事件的理解，并询问是否正确，期盼对方澄清。(I hope you don't mind clarifying this point for me.)

- 这类电子邮件的目的是向对方询问先前不清楚之处，以免造成彼此的误会，因此内容可包括：
  1. 先前的谈话内容。
  2. 不清楚之处。
  3. 向对方道谢。
- 若澄清的事情你都明白了，下次回信时就应该用下列句子："I can see what you mean."（我能理解您的意思。）或者"This is all quite clear to me now."（现在我都很清楚了。）

Sentence Patterns 写作句型 47

损益平衡

I'm afraid I didn't understand what you meant by...
恐怕我不太明白你指的……是什么意思。

- I'm afraid I didn't understand what you meant by "putting on the pressure".
  恐怕我不太明白你指的“施加压力”是什么意思。

Unfortunately, I'm not quite clear on...
不幸的是，我不太清楚关于……

- Unfortunately, I'm not quite clear on the details of the agenda.
  不幸的是，我不太清楚议程的细节部分。

要求重复说明

I was wondering if you could explain (sth) to me again.
不知道你是否可以给我再解释一次（某事）。

- I was wondering if you could explain the instructions to me again.
  不知道你是否可以给我把那些说明再解释一次。

I'd like clarification on... 我想澄清关于……

- I'd like clarification on the strategy you think we should use.
  关于你认为我们所该使用的策略，我想加以澄清。
  * clarification [ˌklærəfəˋkeʃən] *n.* 澄清，说明

要求得知更多细节

Could you please elaborate on...? 能否请你详细说明关于……?

- Could you please elaborate on the health benefits you mentioned?
  关于你提及的健康福利，能请你详细说明吗?
  * elaborate [ɪˋlæbəˌret] *vi.* 详细描述（与介词 on 连用）

Is it possible to be more specific about...?
关于……是否可以说得更具体一些?

- Is it possible for you to be more specific about what was said in the meeting?
  关于会议上所说的事，您是否可以说得更具体一些?

要求举例

Can you give me an example of...? 你可以给我举一个……的例子吗?

- Can you give me an example of the problems in the factory?
  关于工厂的内部问题你可以给我举一个例子吗?

Could you further illustrate what you meant by...?
你能进一步说明你所谓……的意思吗?

- Could you further illustrate what you meant by "unnecessary delays"?
  你能进一步说明你所谓“不必要延迟”的意思吗?
  * illustrate [ˋɪləstret] *vt.* 说明，解释

改变措词

So, what you mean is (that)...? 那么，你的意思是……？

- So, what you mean is (that) we have to reduce shipping costs?
  那么，你的意思是我们必须降低运输成本吗？

If I understand you correctly,... 如果我没有误解的话，……

- If I understand you correctly, the new policy will take effect soon.
  如果我没有误解的话，新政策将很快生效。
  * take effect 生效，开始起作用

Model E-mail 电子邮件范例 47

To: Trevor Miller

From: Charles Wu

Subject: Questions about the promotion

Dear Mr. Miller:

It was wonderful to meet with you yesterday to discuss the **upcoming**[1] promotion. However, there is something I am unclear about. I'm afraid I didn't understand what you meant by "doing everything we can to take the **momentum**[2] away from the competition." I was wondering if you could explain that **concept**[3] to me again. Is it possible to be more **specific**[4] about that?
I hope you don't mind **clarifying**[5] this point for me. I appreciate your taking the time to assist me in this way.

Sincerely,
Charles Wu

中译

收件人：特雷弗·米勒
发件人：查尔斯·吴
主　题：关于促销活动的问题

亲爱的米勒先生：

昨天能与您见面讨论即将举办的促销活动真是太棒了。然而，仍有些事我不太清楚。恐怕我不太明白您所谓“竭尽所能挫一挫竞争者的势头”是什么意思。不知道您是否能再给我解释一次那个概念。是否能再说得更具体些呢？

希望您不要介意为我澄清这一点。我感谢您能抽出时间来如此帮我。

查尔斯·吴　敬上

To: Francis Kelvington

From: Amanda Cai

Subject: Deadline **query**[6]

Hi Francis:

Regarding our telephone conversation this morning, unfortunately, I'm not quite clear on one point. I'd like clarification on the deadline for the first stage of the project. If I understand you correctly, the first stage should begin before the end of May and be finished sometime in November. Is that correct?

I appreciate your help on this—thanks!

Best regards,
Amanda

中译

收件人：弗朗西斯·卡尔文顿
发件人：阿曼达·蔡
主　题：询问截止日期

你好，弗朗西斯：

关于我们今天早上电话中的交谈，不幸的是，有一点我不太清楚。我想澄清这个项目第一阶段的截止日期。如果我没有误解的话，第一阶段应该在5月底之前开始且在11月某个时候结束。这样对吗？

我很感激你对此事的帮助——谢谢！

诚挚的祝福

阿曼达

Vocabulary and Phrases

1. **upcoming** [ˈʌpˌkʌmɪŋ] *a.* 即将来临的
   the upcoming presidential election
   即将举行的总统选举
2. **momentum** [moˈmɛntəm] *n.* 势头，动力
   momentous [moˈmɛntəs] *a.* 关键的，重要的
   gain / gather momentum 势头增大
   a momentous decision / event / occasion
   重大决定/事件/时刻
   例: The anti-corruption campaign is rapidly gaining momentum.
   （反贪污运动的势头正快速增大。）
   The Lakers began to lose momentum in the second half of the game.
   （湖人队在比赛的下半场势头就逐渐减弱。）
3. **concept** [ˈkɑnsɛpt] *n.* 概念
   例: Gary can't seem to grasp the basic concepts of mathematics.
   （加里似乎无法掌握数学的基本概念。）
4. **specific** [spəˈsɪfɪk] *a.* 明确的，具体的
   例: How can you expect me to start if you don't give me specific instructions?
   （如果你不给我明确的指示，怎么能期望我开始做？）
5. **clarify** [ˈklærəˌfaɪ] *vt.* 阐明
   clarification [ˌklærəfəˈkeʃən] *n.* 澄清，说明
   clarify a situation / problem
   澄清情况/问题
6. **query** [ˈkwɛrɪ] *n.* 询问，查询
   例: Should you have a query about our product lines, contact our hot line.
   （若对本公司的产品系列有疑问，请拨打我们的咨询热线。）

Business Writing Exercises

请按括号中的提示将下列句子译成英文。

1. 恐怕我不太了解你们对于全球化的立场。(your stance on globalization)
   ____
2. 不知道你是否可以把你们的营销策略再解释一遍？(I was wondering if you could...?)
   ____
3. 你能否针对你想到的那些点子做详细说明？(Could you please elaborate on...?)
   ____
4. 关于那些销售数字(sales figures)，你是否能说得更具体些？
   ____
5. 如果我的理解无误的话，我们将很快成为贸易伙伴。(trading partners)
   ____

Unit 2

Providing Clarification　给予澄清

Basic Structure　基本结构

1. 点出对方针对前一封邮件已掌握要点。（Your interpretation regarding...is correct.）
2. 针对对方误解之处给予澄清，并详述确切解决之道。（I need to clarify another thing that you mentioned.）
3. 如果对方需要进一步询问，表达愿意解释的意愿。（Please don't hesitate to check with me if you have any further questions.）

- 这类电子邮件的目的是向对方澄清彼此认知上的差距，常见的内容及其他重要句型包括：
  1. 先前误会之处。
     Your understanding / interpretation of... is incorrect.
  2. 做出澄清。
     Allow me to explain...in more detail. / Allow me to elaborate on...
  3. 询问说明是否清楚。
     I hope I have made myself clear on...

Sentence Patterns 写作句型 48

(Sb's) interpretation regarding (sth) is correct.
（某人）对于（某事）的解读是正确的。

- Yes, your interpretation regarding the details of the project is correct.
  没错，你对于这个项目细节的解读是正确的。
  * interpretation [ɪnˌtɜprɪˋteʃən] *n.* 理解；解释

That's exactly what I mean / meant...
那正是我所谓……

- That's exactly what I meant about the best way to control costs.
  那正是我所谓控制成本的最佳方法。

Actually, (sb's) understanding of (sth) is incorrect.
其实，（某人）对（某事）的了解是不正确的。

- Actually, your understanding of the matter is incorrect.
  其实，你对那件事的了解是不正确的。

I'm afraid that's not what I mean / meant...
恐怕那不是我所谓……

- I'm afraid that's not what I meant about increasing productivity.
  恐怕那不是我所谓的提高生产效率。

Allow me to explain (sth) in more detail...
请允许我更详细地解释（某事）……

- Allow me to explain the problem in more detail.
  请允许我更详细地解释这个问题。

Allow me to elaborate on...
让我来阐述……

- Allow me to elaborate on my findings.
  让我来阐述我的调查结果。
  * findings [ˋfaɪndɪŋz] *n.* 调查（或研究）的结果（常用复数）

In other words, ... 换句话说，……

- In other words, we must increase our advertising budget.
  换句话说，我们必须增加我们的广告预算。

To put it another way, ... 换句话说，……

- To put it another way, we have to do our best to keep the costs down.
  换句话说，我们必须尽力压低成本。
  * keep sth down 抑制某事物的增长

I hope I have made myself / everything clear on...

关于……我希望我已经把一切讲得很清楚了。

- I hope I have made everything clear on what the project will entail.
  有关这个项目会有什么需要，我希望我已经把一切讲得很清楚了。
  * entail [ɪnˋtel] *vt.* 需要；使必要

Does that explain things fully regarding...?

有关……一事，那样的说法把一切都解释清楚了吗？

- Does that explain things fully regarding how I would embark on the project?
  有关我将如何进行这个项目一事，那样的说法把一切都解释清楚了吗？
  * embark on...　　从事，着手（新的或艰难的事情）

Model E-mail 电子邮件范例 48

Dear Josh:

I received your e-mail in which you asked me some questions about the Henderson account. Yes, your **interpretation**[2] regarding the complaints they have made about our service is correct. That's exactly what I meant about their **dissatisfaction**[3] with us.

However, I need to **clarify**[4] another thing that you mentioned. Actually, your understanding of what I had in mind for **solutions**[5] to the problem is incorrect. Allow me to explain my ideas on this **in** more **detail**[6]. I don't mean that we should offer any **rebates**[7] or **refunds**[8] to Henderson. **Rather**[9], we should tell them we are willing to give them a 25-percent discount on any future work we do for them. In other words, any type of **compensation**[10] is **contingent**[11] upon them doing further business with us.

I hope I have made my meaning clear on this matter. Please don't hesitate to check with me if you have any further questions.

Regards,
Russ

中译

收件人：乔什·坤
发件人：罗斯·拜尔斯
主　题：客户亨德森

亲爱的乔什：

我收到你的电子邮件，信中你询问我一些关于客户亨德森的问题。没错，有关他们对本公司所提供服务的投诉，你的理解是正确的。那正是我所谓他们对我们不满之处。

然而，我需要澄清你所提到的另一件事。其实，你对我解决这个问题的想法的了解是不正确的。请允许我针对此事更详细地解释我的想法。我的意思不是说我们应该提供任何现金折扣或退款给亨德森。更确切地说，我们应该告诉他们，在未来为他们提供的任何服务，我们愿意提供 75 折的优惠。换句话说，任何形式的补偿要视他们是否继续和我们做生意而定。

我希望针对此事我已经把意思说得很明白了。如果你还有进一步的问题，请别犹豫和我联系。

诚挚的祝福

罗斯

Vocabulary and Phrases

1. **account** [əˋkaʊnt] *n.*（公司，非个人）客户；（银行）账户
   open / close an account
   开立/结清账户
   例：Because of your carelessness, we lost several of our important accounts.
   （因为你的粗心，我们失去了几家重要客户。）

2. **interpretation** [ɪnˌtɝprɪˋteʃən] *n.* 理解；解释
   例：The rules are vague and open to interpretation.
   （这些规定很模糊，可以有各种解释。）

3. **dissatisfaction** [dɪsˌsætɪsˋfækʃən] *n.* 不满（与介词 with 连用）
   例：Jamie expressed her dissatisfaction with the arrangement.
   （杰米表达了她对那项安排的不满。）

4. **clarify** [ˋklærəˌfaɪ] *vt.* 阐明；澄清
   例：I was asked to clarify my position at the meeting.
   （我被要求在会议上阐明我的立场。）

5. **solution** [səˋluʃən] *n.* 解决办法（与介词 to 连用）
   例：There's no simple solution to this problem.
   （这个问题没有简单的解决办法。）

6. **in detail** 详细地
   = at length
   例：We have already discussed this matter in great detail.
   = We have already discussed this matter at great length.
   （我们已经十分详尽地讨论了这个问题。）

7. **rebate** [ˋribet] *n.* 现金折扣
   a cash rebate　现金折扣

8. **refund** [ˋrifʌnd] *n.* 退款
   claim / demand / accept a refund
   要求/要求/接受退款
   例：If you're not satisfied with our service, you can demand a full refund.
   （如果你不满意我们的服务，你可以要求全额退款。）

9. **rather** [ˋræðɚ] *adv.* 更确切地讲；反而（本文中置于句首，用以修饰全句）

例：I woke up late this morning. So, I had to walk, or rather, run to the office.
（我早上起床起晚了。所以，我得走路，更确切地说是跑着去办公室。）

10. **compensation** [ˏkɑmpɛnˋseʃən] *n.* 赔偿（金）

claim / seek compensation
要求赔偿

in compensation for... 以赔偿……

例：The worker received $30,000 in compensation for his workplace injury.
（那位工人获得 3 万美元以作为他工伤的赔偿。）

11. **contingent** [kənˋtɪndʒənt] *a.* 依情况而定的

be contingent upon / on...
视……情况而定

例：Outdoor activities are, as ever, contingent on the weather.
（和往常一样，户外活动要视天气而定。）

* as ever 一如既往

Business Writing Exercises

请按括号中的提示将下列句子译成英文。

1. 关于经济疲软的原因，你的解读是正确的。（the cause of the weak economy）

____

2. 恐怕那并非我所谓削减预算的原意。（reducing budgets）

____

3. 容我更详细地解释我个人的观点。（personal views）

____

4. 换句话说，我们应该设法开源节流。（create more revenue and reduce costs）

____

5. 我希望关于此事我已经把话讲清楚了。

____

Unit 1

Technology and Products 技术与产品

Basic Structure 基本结构

1. 表示参考过各提案后，已选定某种机型。(After examining the various options, I think the best choice is...)
2. 具体说明机型的优点（如高规格制作、完全符合需求、能提高生产效率）。(The machine suits our needs quite well.)
3. 附上完整评估表以作参考。(Please see the attachment for a more complete assessment of...)

- 这类电子邮件的目的是让对方了解选定某机型的理由，因此内容可包括：
  1. 选定的机型。
  2. 选定的理由。
- 如要说明机器的特点（feature）和功能（function），可用“(sth) is equipped with...”的句型说明。
- 一般而言，假如你需要发送附件，不必详述其内容，只要说明附件的性质并要收件人注意就可以了。

Sentence Patterns 写作句型 49

(Sth) is at the cutting edge of...technology. （某物）处在……技术发展的最尖端。

- The new smartphone is at the cutting edge of mobile communications technology.
  这款新的智能手机处在移动通讯技术发展的最尖端。
  * the cutting edge （某事物发展的）最尖端，最前沿

(Sth) is considered state-of-the-art. （某物）被认为是最先进的科技。

- The control panel uses all the newest technology and is considered state-of-the-art.
  这个控制面板使用所有最新的技术，且被认为是最先进的科技。
  * panel [ˋpænḷ] *n.* （汽车或其他机械的）控制板
  state-of-the-art （科技、机电等产品）最先进的

(Sth) is the most advanced type of...available. （某物）是现有的……中最先进的款式。

- The Model SX-78 is the most advanced type of engine available.
  型号 SX-78 是现有的引擎中最先进的款式。

(Sth) is becoming more sophisticated all the time. （某物）变得日益精密。

- Medical techniques are becoming more sophisticated all the time.
  医疗技术变得日益精湛。
  * sophisticated [səˋfɪstɪˏketɪd] *a.* （机器、体系）复杂的，精密的

(Sth) is specifically / specially designed for (sth)
（某物）是特别设计来（做某事）的。

- The method is specifically designed for use in small groups.
  这个方法是专为小组活动设计的。

(Sth) is equipped with (sth) with a capacity of...
（某物）配备有能产生……功率的（某物）。

- The machine is equipped with an engine with a capacity of 1,600 kilowatts.
  这台机器配备有能产生 1600 千瓦功率的引擎。
  * capacity [kəˋpæsətɪ] *n.* 功率；生产能力

(Sth's) configuration allows... （某物的）配置可以……

- The device's configuration allows you to perform many complicated tasks at the same time.
  这个装置的配置让你能同时执行多项复杂的任务。
  * configuration [kənˏfɪɡjəˋreʃən] *n.* 配置，结构

(Sth) is configured to... （某物）被设定为……

- The computer is configured to accept many different types of programs.
  这台计算机被设定为可接受许多不同类型的程序。
  * configure [kənˋfɪɡjɚ] *vt.* 对（设备或软件进行）设定

(Sth) is made to the highest specifications.
（某物）是以最高规格制成的。

- These components are made to the highest specifications.
  这些零件是以最高规格制成的。
  * specification [ˌspɛsəfəˋkeʃən] *n.* 规格（= spec）

If you examine (sth's) specs, you will find...
如果你检查（某物的）规格，你会发现……

- If you examine the product's specs, you will find it meets your requirements.
  如果你检查这个产品的规格，你会发现这项产品符合你的需求。

Model E-mail 电子邮件范例 49

To: Simon Ellingsworth

From: Vincent Gu

Subject: Replacement machinery

Dear Mr. Ellingsworth:

After examining the various options for replacing our aging **equipment**[1], I think the best choice is the proposal **submitted**[2] by Optimum Electronics. Their **machinery**[3] is at **the cutting edge**[4] of technology. Particularly, their Model ABT-962 is very advanced and suits our needs quite well. It is designed to run 24 hours a day and has the **capacity**[5] to handle the **workloads**[6] we require.

Moreover, that model is **configured**[7] to **allow for**[8] the **maximum**[9] productivity with the minimum waste. As the sales reps said, it's made to the highest **specifications**[10]. If you examine the ABT-962's specifications, you will find that it is an excellent piece of equipment.

Please see the attachment for a more complete **assessment**[11] of Optimum Electronics' Model ABT-962.

Best regards,
Vincent

中译

收件人：西门 · 埃林斯沃思

发件人：文森特 · 古

主　题：更换机器

亲爱的埃林斯沃思先生：

审视过各种更换本公司老旧设备的方案后，我认为最好的选择是“绝佳电子产品公司”的提案。该公司的机器处在技术的尖端。尤其是该公司型号 ABT-962 的机器非常先进，且相当符合本公司的需求。那台机器被设计为可全天候运转，并能处理我们需要的工作量。

再者，该型号产品被设定以最低耗损达到最高生产力。正如该公司的销售代表所述，该产品是以最高规格制成的。如果你检查 ABT-962 的规格，就会发现该产品真是一台很棒的设备。

请参阅附件以获得“绝佳电子公司”型号 ABT–962 产品更完整的评估。

诚挚的祝福

文森特

Vocabulary and Phrases

1. **equipment** [ɪˋkwɪpmənt] *n.* 器材(集合名词, 不可数)
   an equipment (×)
   → a piece of equipment　一件器材 (○)
2. **submit** [səbˋmɪt] *vt.* 呈递
   例: The developer submitted the building plan to the committee for approval.
   (开发商将该建造方案呈递给委员会批准。)
3. **machinery** [məˋʃinərɪ] *n.* 机械(集合名词, 不可数)
   machine [məˋʃin] *n.* 机器 (可数)
   a piece of machinery　一台机器
   = a machine
4. **the cutting edge**　技术尖端
   at the cutting edge of computer technology　处在电脑技术的尖端
5. **capacity** [kəˋpæsətɪ] *n.* 功率; 生产能力
   at full capacity　以全力
   例: Since we receive a lot of orders, the factory is now working <u>at full capacity</u>.
   (因为我们收到很多订单, 工厂目前正开足马力生产。)
6. **workload** [ˋwɜkˏlod] *n.* 工作量; 工作负担
   a heavy / light workload
   巨大 / 轻松的工作量
   例: We took on an extra five staff members to cope with the <u>heavy workload</u>.
   (我们额外雇用了 5 名员工来应付这巨大的工作量。)
7. **configure** [kənˋfɪgjɚ] *vt.* 对(设备或软件进行)设定
8. **allow for sth**　使某事有可能
   例: The facts allow for only one explanation.
   (这些事实只可能有一种解释。)
9. **maximum** [ˋmæksəməm] *a.* 最高的 & *n.* 最大量
   minimum [ˋmɪnəməm] *a.* 最低的 & *n.* 最小量
   例: The job requires you to use all your

skills to the maximum.
（这份工作需要你把所有技能发挥到极致。）

The course needs a minimum of ten students to continue.
（这门课最少需要 10 个学生才能继续下去。）

10. **specification** [ˌspɛsəfəˈkeʃən] *n.* 规格；规范（常用复数）

例: The product was built exactly to our specifications.
（这个产品完全是按照我们的规格打造的。）

11. **assessment** [əˈsɛsmənt] *n.* 评价；评估
assess [əˈsɛs] *vt.* 评价

例: What is your assessment of the situation?
（你对这个情况的评估是什么？）

It's difficult to assess the effects of these changes.
（这些改变的效果难以评价。）

Business Writing Exercises

请按括号中的提示将下列句子译成英文。

1. 这些战机（war planes）目前处在航空技术（aeronautic technology）发展的最尖端。

____

2. 微创手术（Minimally invasive surgery）是现有最先进的手术方式。

____

3. 数码相机变得越来越精密。

____

4. 这款手机是专为盲人设计的。

____

5. 如今，几乎每一款手机都配备有内置的数码相机（a built-in digital camera）。

____

Unit 2

Technical Problems
技术上的问题

Basic Structure 基本结构

1. 表明来信是要提及某技术上的问题。（I thought it was best to send you an e-mail about a technical problem.）
2. 明确说明问题所在。（There was a problem detected in the machine.）
3. 表示解决问题至关重要。（It's crucial that we work hard to solve this problem.）

- 这类电子邮件的目的是让对方了解技术上所发生的问题，因此内容可包括：
  1. 说明问题可能出在哪里。
  2. 说明此问题造成的影响。

Sentence Patterns 写作句型 50

There has been / was a bug detected somewhere in... 在……中某处检测出毛病。

- As soon as I turned on my computer, there was a bug detected somewhere in the database.
  我一打开电脑，在数据库某处就检测出毛病。
  * bug [bʌg] *n.* （机器，尤指电脑的）毛病

(Sb / Sth) is experiencing a glitch. （某物）遇到小故障。

- My computer is experiencing a glitch that is preventing me from logging on the Facebook.
  我的电脑遇到小故障，让我无法登录脸书。
  * glitch [glɪtʃ] *n.* 小故障；小毛病

(Sth) has been / was replaced after a defect in...was discovered.
……内被发现有缺陷后（某物）就被替换掉了。

- All model RT-124 machinery has been replaced after a defect in the engine was discovered.
  所有型号 RT-124 的机器在被发现引擎有缺陷后都被替换掉了。
  * defect [ˈdifɛkt] *n.* 缺陷

(Sth) has been / was found to be defective. （某物）有瑕疵。

- This piece of equipment was found to be slightly defective.
  这台设备被发现稍微有点瑕疵。
  * defective [dɪˈfɛktɪv] *a.* 有瑕疵的

report a malfunction in / of... 通报……的故障

- Shortly before the crash, the pilot reported a malfunction in one of the engines.
  在飞机失事坠毁前不久，飞行员报告其中一台引擎发生故障。
  * malfunction [ˌmælˈfʌŋkʃən] *vi.* （机器）运转失常；出现故障

We are experiencing technical problems with (sth).
我们现在正遇到（某物）技术上的问题。

- Right now, we are experiencing technical problems with the scanner.
  现在我们遇到扫描仪的一些技术问题。

(Sth) is incompatible / not compatible with (sth). （某物）和（某物）不兼容。

- We have to make sure that the new system is compatible with the existing equipment.
  我们必须确保新系统可与现有的设备兼容。
  * compatible [kəmˈpætəbḷ] *a.* （机器、电脑）兼容的

(Sth) does not recognize (sth). （某物）无法识别（某物）。

- My notebook doesn't recognize this new software.
  我的笔记本电脑无法识别这套新软件。

(Sth) is lost in cyberspace. （某物）遗失在网络空间。

- I believe the e-mail I sent you is lost in cyberspace somewhere.
  我认为我发给你的那封电子邮件遗失在网络空间了。

There is a problem with Internet connectivity. 网络连接上出了问题。

- There is a problem with Internet connectivity at my office.
  我办公室里的网络连接出了问题。

Model E-mail 电子邮件范例 50

To: Mason Liang, Project Manager

From: Arthur Collier, Senior IT Specialist

Subject: IT problems

Dear Mr. Liang:

I thought it was best to send you an e-mail about a technical problem we've been having recently. There is a bug in the new software that R&D is currently developing. I strongly **suspect**[1] that **the source code**[2] has a **defect**[3]. I'm not exactly sure what the problem is, but we are experiencing many technical problems with the program.

In addition, the software is not **compatible**[4] with a few other common programs on the market. For example, it doesn't **recognize**[5] some Mac **applications**[6].

There is one more thing I'd like to bring to your attention. After the **installation**[7] of the software, some of our workers reported a **malfunction**[8] in their computers. It's **crucial**[9] that we work hard to solve these problems.

I sent you an e-mail last week about these problems, but I think it got lost in cyberspace and never reached you.

Yours truly,
Arthur

中译

收件人：梅森·梁
项目经理
发件人：亚瑟·科立尔
高级 IT 专家
主　题：IT 问题

亲爱的梁先生：

关于我们最近一直遇到的一个技术问题，我想我最好给你发一封电子邮件。研发部目前正在研发的新软件里有病毒。我强烈怀疑原始码有缺陷。我不太确定是什么问题，但我们使用此程序时遇到了许多技术上的问题。

此外，该软件和市面上一些其他常见程序不兼容。举例来说，该软件无法识别一些苹果计算机的应用程序。

还有一件事我想提请你注意。安装好这个软件后，我们的一些工作人员呈报计算机发生故障。我们要努力解决这些问题，此事至关重要。

为了这些问题我上星期给你发了一封电子邮件，但我想那封邮件应该是消失在网络空间了，从未到你手中吧。

亚瑟　敬上

Vocabulary and Phrases

1. **suspect** [sə'spɛkt] *vt.* 怀疑

例: If you suspect a gas leak, do not strike a match.
（假如你怀疑有煤气泄漏，就不要划火柴。）

You have no reason to suspect that they are trying to get rid of you.
（你没有理由怀疑他们试图要摆脱你。）

2. **the source code** 原始码

3. **defect** ['difɛkt] *n.* 缺陷；毛病

例: There are so many defects in our current system.
（我们现行系统有太多毛病了。）

4. **compatible** [kəm'pætəbḷ] *a.* 兼容的
be compatible with... 与……兼容

例: The new software is not compatible with older operating systems.
（新软件和旧的操作系统不兼容。）

5. **recognize** ['rɛkəgˌnaɪz] *vt.* 识别出

例: Do you recognize this tune?
（你能听出这是哪首曲子吗？）

6. **application** [ˌæplə'keʃən] *n.* 应用程序；应用软件

7. **installation** [ˌɪnstə'leʃən] *n.* 安装
install [ɪn'stɔl] *vt.* 安装

例: My rough estimation is that the installation of the new system will take several days.
（我粗略估计这个新系统的安装要花几天的时间。）

I had difficulty installing this software.
（我在安装这个软件时遇到了麻烦。）

8. **malfunction** [mæl'fʌŋkʃən] *n. & vi.* 机器失灵

9. **crucial** ['kruʃəl] *a.* 重要的

例: It's crucial that we get the priorities straight.
（我们分清事情的轻重缓急很重要。）

Winning the contract is crucial to our company.
（赢得那份合约对我们公司来说很重要。）

Business Writing Exercises

请按括号中的提示将下列句子译成英文。

1. 硬盘上已检测出有问题。

______

2. 目前我们遇到软件方面的小故障。（a software glitch）

______

3. 这个设计已被发现有瑕疵。

______

4. 该飞行员已报告其飞机导航系统（navigation system）发生故障。

______

5. 该软件跟我的操作系统（operating system）不兼容。

______

Chapter 1 Unit 1

1 The reason I am writing to you is to thank you for all that you have done for me.

2 I joined the company in the summer of 2008.

3 Previously, I worked part-time at a convenience store.（也可写成：I was a part-time worker at a convenience store.）

4 Now, I am responsible for quality control of our products.

5 I look forward to cooperating with you in the future.

Chapter 1 Unit 2

1 Our company specializes in building websites for small businesses.

2 We manufacture all kinds / types of furniture.

3 We offer a full range of kitchen utensils.

4 Our company was founded / established in August, 1997.

5 We are based in Tokyo, but we have a branch office in Taipei.

Chapter 1 Unit 3

1 This shirt comes in three sizes and seven colors.

2 The LCD TV can also function as a monitor.

3 We offer a wide array / selection of clocks and watches.

4 I would like to remind you that all our products are reliable and durable.

5 Don't hesitate to call me if you need any help.

Chapter 2 Unit 1

1 Are you available on Friday evening?

2 What's your schedule like in the next few days?

3 Please send me an e-mail at your earliest convenience.

4 Let me check my calendar before you book a flight for me.

5 I'll get back to you as soon as the meeting is over.

Chapter 2 Unit 2

1 I'm free on Thursday morning but not on Wednesday afternoon.

2 June 8 works for me. How about you?

3 I'm afraid I'm tied up all day tomorrow.

4 Friday at 3 p.m. is preferable to me.

5 With regard to the date of the meeting, do you have any suggestions?

Chapter 2 Unit 3

1 I'm afraid (that) we must take immediate action.

2 Unfortunately, the plane had taken off when I arrived at the airport.

3 Can you reschedule the meeting to next Monday?

4 Due to circumstances beyond our control, we are unable to meet the deadline.

5 Sorry for any inconvenience the delay has caused.

Chapter 2 Unit 4

1 Time is running short, so let's move to the last item on the agenda.

2 Attached is a copy of my résumé.

3 The first issue we will deal with is one of our new production lines.

4 Also on the table for discussion is whether we should make more investments abroad.

5 The meeting is scheduled to last no more than 3 hours.（也可写成：The meeting is scheduled to last 3 hours at (the) most.）

Chapter 3 Unit 1

1 I'm planning a business trip to New York from May 10 to May 17.

2 My flight will arrive at about 10 a.m. if it is not delayed.

3 I would really like to meet with you about our cooperation plans.

4 Would it be OK for you to pick me up at the airport?

5 Hopefully, we can keep in touch / contact more often in the future.

Chapter 3 Unit 2

1 I've booked a room at that hotel in your name.

2 I'm calling to tell you (that) your seat reservation is confirmed.

3 The check-out time in most hotels is noon.

4 Is breakfast included in the price of the room?

5 You can catch a shuttle bus to our hotel if you don't want to take a taxi.

Chapter 3 Unit 3

1 Here is the itinerary for your trip around the island.

2 The flight departs at 5 p.m. from Terminal 2.

3 There is a 2-hour stopover / layover in Taipei.

4 According to your request, you will be picked up from the airport.

5 If you like, we could arrange a direct flight for you.

Chapter 4 Unit 1

1 It's been a long time since we last met.

2 Can you recommend a good Japanese restaurant around here?

3 We really should get together for lunch or coffee sometime.

4 If you were buying a bag, which brand would you choose?

5 What would you suggest regarding further reading?

Chapter 4 Unit 2

1 I suggest Mary (should) take a few days off.

2 Why don't you make a backup copy?

3 Beitou is famous for its hot springs and restaurants.

4 If I were you, I would apologize to your colleague.

5 In terms of the box office, the film / movie is a huge success.

Chapter 5 Unit 1

1 I would like to invite you to be a judge of our speech contest.

2 I was wondering if you are interested in becoming a singer.

3 Do you feel like going swimming on the weekend?

4 I'm afraid I can't make it to your birthday party.

5 We request the pleasure / honor of your presence at our wedding anniversary.

Chapter 5 Unit 2

1 That's kind of you to offer help, but I can manage it myself.

2 I can't stay for dinner, but a cup of tea sounds great.

3 It's necessary for you to exercise every day.

4 I'm really sorry, but I already have other plans.

5 Congratulations on your recent promotion to manager!

Chapter 6 Unit 1

1 I'm writing to inquire about your new products and their prices.

2 Could you send me a catalogue of your electrical appliances?

3 I would appreciate it if you could keep me informed / posted.

4 We are a furniture manufacturer (that is) based in Malaysia.

5 We are in the market for 30 Christmas trees.

Chapter 6 Unit 2

1 Could you provide me with a quotation for 1,000 umbrellas?

2 Please quote us a price for the following items.

3 What's your best price for 500 pairs of jeans?

4 Do you offer any discounts on bulk / sizeable orders?

5 What is the minimum order for delivery?

Chapter 6 Unit 3

1 Please find our price list attached to this e-mail.

2 We are pleased to provide you with our newest catalog.

3 Due to price fluctuations, the quote is valid for 3 days.

4 This quotation is based on a minimum order of 600 units.

5 If any problem comes up, please don't hesitate to contact us.

Chapter 7 Unit 1

1 I'd like to place an order for 2,000 ballpoint pens and 60 staplers.

2 Could you please confirm that the price is correct and get back to us?

3 As agreed to previously, I believe the amount is wrong.

4 Please ship our order to the following address.

5 Please ensure delivery by March 15. Thanks!

Chapter 7 Unit 2

1. Thank you for your order for 100 sets of kitchenware.
2. We are writing to confirm your order.
3. We are currently in the process of filling your order.
4. The goods are expected to reach you by March 15.
5. Your business with us is very much appreciated and valued by us.

Chapter 7 Unit 3

1. Unfortunately, we are unable to fill your order as requested.
2. The mountain bikes you ordered are currently out of stock.
3. I'm sorry that the model you ordered has been discontinued.
4. Due to circumstances beyond our control, we have to cancel our order.
5. Please bear with us. We'll make up for your losses.

Chapter 8 Unit 1

1. Our records show that your payment is long overdue.
2. We request that you settle the invoice within a week.
3. If the payment has already been made, please disregard the notice.
4. If you have any questions, please contact us as soon as possible.
5. Unless sound reasons are given, we will be forced to take legal action.

Chapter 8 Unit 2

1. Please find enclosed a check for $2,500.
2. The amount of 480 euros has been transferred to your own account.
3. Please confirm that you have received the order we placed with you yesterday.
4. A large amount of information is available on the Internet.
5. A large number of passengers are waiting for the train right now.

Chapter 8 Unit 3

1. This letter is to acknowledge receipt of your payment for 30 desktops.
2. Thank you very much for your check payment dated December 3.
3. We have received notification of the transference of $6,000.
4. This e-mail is to confirm we have received your payment in the form of credit card.
5. The bank record shows our account has been credited with $3,050 for payment of your order.

Chapter 9 Unit 1

1. I'm writing to complain about your shipping service.
2. There is a problem with the goods you delivered this morning.
3. I regret to inform you that the compact discs we received were damaged.

4 In addition, there are inconsistencies in quality that need to be addressed.

5 I hope these problems can be solved quickly and efficiently.

Chapter 9 Unit 2

1 We appreciate being informed of the defects of our products.

2 Please accept our sincerest apologies for the delay in delivery.

3 The problem was due to mechanical failure rather than human negligence.

4 Please rest assured that we are doing everything possible to rectify the problem.

5 We are sorry for any inconvenience we may have caused.

Chapter 10 Unit 1

1 What is the current status of the job market in our country?

2 Have you made much progress on your plans of cooperation?

3 I hope everything is going smoothly in the establishment of another production line.

4 Will you be able to finish the work by the deadline?

5 I'm getting a bit worried about our overseas investments.

Chapter 10 Unit 2

1 So far, we haven't encountered any major problems with our equipment.

2 Please rest assured that everything is going according to schedule.

3 However, there is a slight problem with our air conditioning.

4 Due to a major strike, we have fallen behind schedule.

5 As a result, I'm afraid we can't meet your deadline.

Chapter 11 Unit 1

1 What I propose is to find a famous person to endorse our new product.

2 Allow me to point out that the time is not right for making important decisions.

3 From my point of view, some of our equipment needs to be replaced or updated.

4 The way I see it, the recession will last longer than we thought.

5 As far as I'm concerned, I don't think your idea can work.

Chapter 11 Unit 2

1 Thank you for submitting your comments regarding our customer service.

2 I'll pass along your suggestions to our manager.

3 Some new company policies are currently under evaluation.

4 I'm sure (that) your proposal will be taken into consideration.

5 The CEO has the final say on almost everything.

Chapter 12 Unit 1

1 These machines are maintained every three months.

2 More than 50 employees were laid

off last year.

3 A new album will be released in the middle of June.

4 A teleconferencing system has been installed in our office.

5 The inventory work is being carried out / conducted right now.

Chapter 12 Unit 2

1 Most people believe the rising prices are due to food shortages.

2 As a result of our joint efforts, our sales have increased a lot.

3 Mr. Wang retired early due to poor health.

4 They kept losing money. Consequently, they had to close down their business.

5 The weakness of the US dollar sparked an international competition for gold.

Chapter 12 Unit 3

1 Similarly, prices go up when demand increases.

2 On the other hand, prices go down when supply is greater than demand.

3 For example, we may recycle resources to protect the environment.

4 As a matter of fact, we are expanding our business now.

5 In other words, we will need to hire more staff.

Chapter 12 Unit 4

1 On the whole, most passengers are satisfied with our service.

2 At any rate, we have gone out of the red.

3 In summary, we need to persist to achieve success.

4 In short, you have to size up the situation before making an investment.

5 In conclusion, customers are always right.

Chapter 13 Unit 1

1 It is kind of you to offer me such a great opportunity.

2 Your hospitality was greatly appreciated by all of us.

3 It was thoughtful of you to give me a ride to the airport.

4 I can't thank you enough for the time and effort you put in.

5 I'm more than happy to stay and keep you company.

Chapter 13 Unit 2

1 I wish to extend my heartfelt congratulations on your 50th wedding anniversary.

2 I was so happy to hear that you were going to get married.

3 It was terrible to learn that your company had closed down.

4 Please accept my condolences on the death of your father.

5 Wishing you a pleasant weekend!

Chapter 14 Unit 1

1 I'm writing to apply for the position of marketing manager with your company.

2 I have five years of experience working as a secretary.

3 I am an optimistic and aggressive person.

4 My career goal is to become a famous lawyer.

5 My job duties include keeping in contact with old customers and looking for new ones.

Chapter 14 Unit 2

1 Thank you for taking the time to read my report.

2 I'm quite interested in working for your company as a sales representative.

3 I wish to reiterate that I am a responsible and efficient person.

4 That interview has reinforced my belief for setting up my own business.

5 Once again, I'd like to thank you for your kind invitation.

Chapter 15 Unit 1

1 We are prepared to offer you a price of $10 per unit.

2 How would you feel about an offer of $15,000?

3 The best price we can offer is $105.

4 What do you propose regarding the total price?

5 What do you have in mind regarding our bulk order?

Chapter 15 Unit 2

1 We can lower the price on condition that you pay in cash.

2 We insist on an upfront payment of 10%.

3 If you give us a 20% discount, we will increase our order.

4 Would you accept the unit price of $12.6?

5 Please let us know if these terms are acceptable to you.

Chapter 15 Unit 3

1 We are willing to accept your terms.

2 I think we can go along with what you are offering.

3 I'll get back to you on the price as soon as possible.

4 We need more time to consider your offer.

5 Your estimate isn't even in the ballpark.

Chapter 16 Unit 1

1 Revenues are forecasted to amount to $5 million.

2 Our expenditures are expected to be $2 million.

3 The estimated deficit is expected to reach $10 million dollars.

4 Our surplus is expected to be more than $1 million.

5 We expect revenues to exceed expenditures by $3 million.

Chapter 16 Unit 2

1 On average, our company makes a profit of $5 million a year.

2 They suffered a loss of $10 million in the third quarter.

3 Our company neither made a profit nor a loss last year.

4 They made a killing selling organic vegetables and fruit.

5 A lot of people lose their shirts in the stock market.

Chapter 17 Unit 1

1 I'm sure we are able to meet our sales objectives next year.

2 Last month's sales volume fell short of our expectations.

3 The sales turnover for the third quarter was $6 million.

4 The sales forecast for the next quarter has been completed.

5 We need to set sales quotas for every sales representative.

Chapter 17 Unit 2

1 We are going to launch a sales campaign in November.

2 We need to create a buzz about our new products.

3 They are offering cash rebates on electric appliances.

4 They are giving away free samples of cosmetics.

5 The two department stores are engaging in a price war.

Chapter 18 Unit 1

1 First of all, we must be able to define our target market.

2 Research and development will help us gain more market share.

3 Teenage consumers make up a large segment of the market.

4 Fake goods / items have flooded the market recently.

5 Smartphones have been in great demand since last year.

Chapter 18 Unit 2

1 After many years' effort, we have become a global brand.

2 They would like to establish brand loyalty among customers as soon as possible.

3 Can you differentiate the real thing from the fake?

4 I strongly recommend using blogs to promote sales.

5 Brand-name products are more durable than generic products.

Chapter 19 Unit 1

1 I'm afraid I don't (quite) understand your stance on globalization.

2 I was wondering if you could explain your marketing strategies again?

3 Could you please elaborate on those ideas that you came up with?

4 Is it possible for you to be more specific about those sales figures?

5 If I understand you correctly, we are going to become trading partners soon.

Chapter 19 Unit 2

1 Your interpretation regarding the cause of the weak economy is correct.

2 I'm afraid that's not what I meant about reducing budgets.

3 Allow me to explain my personal views in more detail.

4 To put it another way, we should try to create more revenue and reduce costs.

5 I hope I've made myself clear on this matter.

Chapter 20 Unit 1

1 These war planes are at the cutting edge of aeronautic technology.

2 Minimally invasive surgery is the most advanced type of surgery available.

3 Digital cameras are becoming more sophisticated all the time.

4 This cell phone is specially designed for blind people.

5 Nowadays almost every cell phone is equipped with a built-in digital camera.

Chapter 20 Unit 2

1 There has been a bug detected on the hard drive.

2 We are now experiencing a software glitch.

3 The design has been found to be defective.

4 The pilot has reported a malfunction in the aircraft's navigation system.

5 The software is incompatible with my operating system.

英式 IPA 音标与美式 K.K. 音标对照表

序号	IPA	K.K.	Key Words
1	ɪ	ɪ	**bit**
2	e	ɛ	**bed**
3	æ	æ	**cat**
4	ɒ	ɑ	**hot**
5	ʌ	ʌ	**cut**
6	ʊ	ʊ	**put**
7	ə	ə	**about**
8	i	ɪ	**happy**
9	u	ʊ	**actuality**
10	iː	i	**bee**
11	ɑː	ɑ	**father**
12	ɔː	ɔ	**law**
13	uː	u	**tool**
14	ɜː	ɝ	**bird**
15	eɪ	e	**name**
16	aɪ	aɪ	**lie**
17	ɔɪ	ɔɪ	**boy**
18	əʊ	o	**no**
19	aʊ	aʊ	**out**
20	ɪə	ɪr	**beer**
21	eə	ɛr	**hair**
22	ʊə	ʊr	**tour**
23	uə	ʊə	**actual**
24	iə	jɚ	**peculiar**

Notes:

1. K.K.音标取自美国两位语言学家 John S. Kenyon 和 Thomas A. Knott 两人姓氏的第一个字母。其特点是按照一般的美国读法标音。
2. 本列表所用的 IPA 音标是英国 Jones 音标的最新修订形式。
3. K.K.音标除辅音中[N]、[L]与IPA音标[n]、[l]符号不同外，其余基本一致。
4. 美式和英式发音的不同点之一是卷舌音。如果单词的字母组合含有 r，该组合一般会发卷舌音。例：four [fɔː]（IPA）； [fɔr]（K.K.）。